UNITED NATIONS CONFERENCE ON TRADE AND DEVELOPMENT

UNCTAD

I0110180

TECHNOLOGY AND INNOVATION REPORT 2015

Fostering Innovation Policies for Industrial Development

UNITED NATIONS
New York and Geneva, 2015

NOTE

The terms country/economy as used in this Report also refer, as appropriate, to territories or areas; the designations employed and the presentation of the material do not imply the expression of any opinion whatsoever on the part of the Secretariat of the United Nations concerning the legal status of any country, territory, city or area or of its authorities, or concerning the delimitation of its frontiers or boundaries. In addition, the designations of country groups are intended solely for statistical or analytical convenience and do not necessarily express a judgment about the stage of development reached by a particular country or area in the development process. The major country groupings used in this Report follow the classification of the United Nations Statistical Office. Details of the classification are provided in Annex I of this Report.

The boundaries and names shown and designations used on the maps presented in this publication do not imply official endorsement or acceptance by the United Nations.

Symbols which may have been used in the tables denote the following:

• Two dots (..) indicate that data are not available or are not separately reported. Rows in tables are omitted in those cases where no data are available for any of the elements in the row.

• A dash (–) indicates that the item is equal to zero or its value is negligible.

• A blank in a table indicates that the item is not applicable, unless otherwise indicated.

• A slash (/) between dates representing years (e.g., 1994/95) indicates a financial year.

• Use of a dash (–) between dates representing years (e.g. 1994–1995) signifies the full period involved, including the beginning and end years.

• Reference to "dollars" ($) means United States dollars, unless otherwise indicated.

• Details and percentages in tables do not necessarily add to totals because of rounding.

The material contained in this study may be freely quoted with appropriate acknowledgement.

This publication has been edited externally.

UNITED NATIONS PUBLICATION

UNCTAD/TIR/2015

Sales No. E.15.II.D.3

ISSN 2076-2917

ISBN 978-92-1-112889-5

e-ISBN 978-92-1-057306-1

PREFACE

Building productive capacities and promoting sustainable industrialization have an important role to play across the spectrum of the integrated 2030 Agenda for Sustainable Development. The Agenda recognizes that the notion of sustainable industrialization is multi-faceted: it is not solely limited to environmental sustainability, but refers to efforts that are technology-led, productivity enhancing and poverty-reducing. It is based on the understanding that no industrial policy is complete without an accompanying innovation policy. Both are essential and complementary to shaping developmental outcomes and creating prosperity for all.

The UNCTAD Technology and Innovation Report of 2015 addresses this urgent policy priority by analyzing the crucial role of technological learning and innovation capacity. Promoting industrialization is a challenge throughout the world. This report helps to address some of the questions that policymakers face when seeking to forge new paths to secure a prosperous future for their people.

I encourage governments, policymakers and development partners to use this report as a resource as they seek to formulate the most effective approaches to achieving the Sustainable Development Goals.

BAN Ki-moon

Secretary General

United Nations

ACKNOWLEDGEMENTS

The Technology and Innovation Report 2015 was prepared by a team comprising Padmashree Gehl Sampath (team leader and main author), Donatus Ayitey and Mesut Saygili under the direction of Anne Miroux, Director of Division on Technology and Logistics, UNCTAD.

This report would not have been possible without the support of national agencies that helped collect primary data in Nigeria, Tanzania and Ethiopia. In Nigeria, the collaboration with Femi Olukesusi (Director and Professor, Policy Engagement and ICT, NISER) and his colleagues at the Nigerian Institute for Social and Economic Research (NISER) is acknowledged for the questionnaire survey and collection of data. Similarly, in Tanzania, UNCTAD is grateful to colleagues at the Tanzanian Commission on Science and Technology who partnered to conduct a national workshop, administer the questionnaire survey and conduct interviews. Hassan Mshinda (Director General, Commission for Science and Technology, Tanzania), Omar Bakari (Coordinator, Commission for Science and Technology's Cluster Development Programme, Tanzania), Flora Tibuwana, and Festo Maro's inputs are particularly acknowledged. In Ethiopia, UNCTAD is grateful to support of Mr. Melkamsew Abate and Mr. Dessie Abeje, Ministry of Trade and Industry.

Comments and suggestions provided by the following experts during the Geneva Peer Review meeting to discuss the outline and methodology are gratefully acknowledged: Bengt-Åke Lundvall (Professor, University of Alborg Denmark), Helena Forsman (Professor in Business Management, University of Tampere, Finland), Mark Nicklas (Deputy Head, European Commission's Innovation Policy and Investment Unit), Christian Berggren (Professor, Industrial Management, Linköping University, Sweden), Biswajit Dhar (Director General, Research and Information Allied Systems, New Delhi), Conrad Von Igel Grisar (Executive Director, InnovaChile, Consejo Nacional de Innovación para la Competitividad, Chile), Flávia Kickinger (Head, Innovation Department , BNDES , Brazil), Ato Eyasu Dessalegne (Director, S&T Policy Research Directorate, MOST, Ethiopia), Omar Bakari (Coordinator, Commission for Science and Technology's Cluster Development Programme, Tanzania), Pedro Roffe (Senior Associate, ICTSD, Geneva, Switzerland) and Kanchana Wanichkorn (Director, Policy and Research Management, National Science Technology and Innovation Policy Office, Thailand). Comments by the following experts on the first draft of the report during a Peer Review Meeting in Addis Ababa, Ethiopia are acknowledged: Prof. Judith Sutz (University of Monte Video), Prof. Susan Cozzens (University of Georgia Tech), Prof. K. J. Joseph (Centre for Development Studies, Trivandrum), Prof. Keun Lee (Seoul Technical University), Prof. Joanna Chataway (Rand Corporation), Bitrina Diyamett (Executive Director, STIPRO, Tanzania).

Country chapters benefited from comments provided by Banji Oyelaran-Oyeyinka (Director, Monitoring and Research, UN-HABITAT, Nairobi, Kenya), and Femi Olokesusi (Director, Policy Engagement and ICT, Nigerian Institute of Social and Economic Research, Nigeria) on the chapter on Nigeria; Hassan Mshinda (Director General, COSTECH), Omar Bakari (COSTECH) and mr. Maduka Kessay, (President's office and Deputy Executive Secretary of the Planning Commission) on the chapter on Tanzania and Carl Daspect (Head of Trade and Economic Section, EU Delegation to Ethiopia, Addis Ababa) and Taffere Tesfachew, Richard Kozul Wright and Ermias Tekeste Biadgleng (UNCTAD) on the chapter on Ethiopia.

Comments from the following colleagues are also acknowledged: Angel Gonzalez-Sanz (Division on Technology and Logistics), Torbjorn Fredriksson (Division on Technology and Logistics), Lisa Borgatti (Division for Africa, Least Developed Countries and Special Programmes), Piergiuseppe Fortunato (Division on Globalization and Development Strategies), and Axele Giroud and Fiorina Mugione (Division of Investment and Enterprise). Inputs from Tansug Ok and Arun Jacob to the Tanzania chapter during the early stages of the preparation of the report are acknowledged. Research assistance by Guiliano Loungo and secretarial assistance by Malou Pasinos in organizing the field missions is also gratefully acknowledged.

The report was edited by Mark Bloch and Sophie Combette was responsible for the cover page. Nathalie Loriot was responsible for the layout.

LIST OF ABBREVIATIONS

AfDB	African Development Bank
AMCOST	African Ministerial Conference on Science and Technology
BIS	Basic Industrialization Strategy
COMESA	Common Market for Eastern and Southern Africa
COSTECH	Tanzania Commission for Science and Technology
CPA	Consolidated Plan of Action
EAC	East African Community
ECOWAS	Economic Community of West African States
EPZ	Export Processing Zone
EXIM	Export-Import Bank
FB-PIDI	Food and Beverages and Pharmaceuticals Industry Development Institute
FDI	Foreign Direct Investment
GDP	Gross Domestic Product
GTP	Growth and Transformation Plan
GURT	Government of the United Republic of Tanzania
ICT	Information and Communication Technology
IDB	Inter-American Development Bank
IIDS	Integrated Industrial Development Strategy
ILO	International Labour Organization
IPR	Intellectual Property Right
LDC	Least Developed Country
MoFED	Ministry of Finance and Economic Development
MoST	Ministry of Science and Technology
MSME	Micro, Small and Medium Enterprise
NBS	National Bureau of Statistics
NEPAD	New Partnership for Africa's Development
NIP	National Industrial Policy
NRDP	National Research and Development Policy
NSGRP	National Strategy for Growth and Poverty Reduction
NV	Nigeria Vision
PASDEP	Plan for Accelerated and Sustained Development to End Poverty
PPP	Public Private Partnership
R&D	Research and Development
SADC	Southern African Development Community
SEZ	Special Economic Zone
SIDP	Sustainable Industrial Development Policy
SME	Small And Medium-Sized Enterprise
STI	Science, Technology and Innovation
TCCIA	Tanzania Chamber of Commerce, Industry and Agriculture
TDV	Tanzania Development Vision
TeCAT	Technology Capability Accumulation and Transfer
WDI	World Development Indicators

CONTENTS

List of figures

List of boxes

List of tables

OVERVIEW

I. INNOVATION AND INDUSTRIAL POLICIES HAVE BOTH REGAINED IMPORTANCE

Industrialization is by no means an easy process. This report is set against the broader international context, wherein a large number of countries have placed renewed emphasis on policy frameworks on industrial policies and science, technology and innovation (STI or innovation policies) to address the challenge of fostering industrialization and closing the technology gap. This report analyses an issue that is of high policy relevance, namely: how can synergies between industrial and innovation policy frameworks be harnessed to help countries to leverage overall growth and transformation.

In the quest to promote a 'great transformation' of sectors and the economy, industrial development and STI policies overlap on the question of promoting technological learning and competence building. These overlaps assume added importance for developing countries as they often lead to a parallel narrative on technological learning. In practice, this implies that the incentives and instruments of both policies are often quite similar; furthermore, they tend to lead to duplication of scarce resource, inter-agency rivalries and less than satisfactory outcomes when they are not accompanied by well-coordinated policy processes.

A second reason why the overlap matters is that both policies approach technological learning from different perspectives. For example, while industrial development strategies set overall economic targets, innovation policies provide the institutional infrastructure for learning, as well as individual targets and supportive incentives to firms. While industrial development strategies aim to develop high-technology sectors, stimulate job growth and eradicate poverty, priority sectors and the *modus operandi* for such prioritization is usually set out in STI frameworks. Similarly, the industrial development strategy of a country may emphasize job growth, particularly to facilitate recovery from the economic and financial crisis of 2007-2008, but it is the STI framework that determines how this job growth can be based on technological development, and how high-quality and sustainable jobs can be created. Despite these overlaps and the complementary nature of both policy frameworks, neither of them is redundant, and close coordination is crucial to enforce developmental outcomes.

While there are some good examples of countries within the developing world that have historically coordinated their industrial development strategies with STI policy objectives, there have also been an equal number of countries that have not managed to do so. Friction has long existed between the two sets of policies due to the fact that consolidation of existing industry (which in many countries is still traditional, or predominantly composed of SMEs), or the promotion of innovation and industrial development are seen as two separate issues.

II. COORDINATING THEIR IMPACT IS ESSENTIAL FOR DEVELOPMENTAL OUTCOMES

Industrial development and innovation are not either/ or options. Industrial upgrading, whether in traditional or new sectors, cannot be achieved without promoting technological upgrading and innovation capacity. The inability to acknowledge and foster this relationship has been the undoing of several developing countries, and has resulted in local industries being unable to enhance productivity despite repeated industrial policy efforts, mainly because there was no emphasis on technological change at the firm level.

Coordinated frameworks on industrial development and technology and innovation capacity need to be emphasized by all countries; a good start in this regard is to understand the links that exist between the two policies and how they impact key actors in the industrialization process, namely, the state, the market, the private and public sectors and domestic and foreign actors. The experiences of East Asian countries and other emerging economies illustrates that getting the right mix of interventions to foster the interaction between these actors is critical for successful industrialization. Crucial questions need to be reframed, and choices refined. For example, it is not whether to foster public research or not, but rather how much public research is needed to boost the local private sector. Similarly, the concern is not whether there should be foreign direct investment (FDI) or not, but rather what is the right kind of FDI, and how can it enhance technology absorption capacity.

Finding the appropriate balance and the 'right' combination of incentives is contingent on how the two policies interact, not just at the policy definition level, where policy goals and targets are set, but also on the mix of incentives contained in these policies, as appropriate to the local context. This rests on how the policies are coordinated, and more specifically with a focus on *getting the policy processes right*. An innovation and industry-friendly climate is therefore not about just specifying/ granting a broad range of incentives, but has rather more to do with identifying the activities, the beneficiaries that need support (i.e. the kind of firms and what they should be focusing on), and how such support can be coordinated through existing agencies. Goal 9 of the 2030 Agenda for Sustainable Development embodies this imperative for coordinating industrial development with fostering innovation. Making strides towards industrial development in years to come will hinge upon identifying and promoting these linkages between innovation and industrial policies from a practical perspective, to avoid pitfalls and channel opportunities for local economies.

In practice, therefore, a synergistic environment for innovation-led industrial de-velopment rests on coordination of policy implementation at the macro-, meso- and micro-levels. At the macro-level (i.e. at the level of national oversight and policymaking), policy frameworks on both industrial development and STI policy should be articulated to provide a lean and cogent conceptualization of common goals and objectives. The coordinated implementation of these policy frameworks occurs at meso-levels, i.e. when the policies are translated into implementation through incentives, programmes and agency mandates. The impact of these policies on firm-level performance occurs at the grassroots level, and is hence a micro-issue, which is affected by a range of factors that impact day-to-day performance. Without coordination at all three levels, it would negatively impact firm-level performance and vitiate the common goal of promoting technology-led industrial growth, even if countries have relevant policy frameworks on industrial development and innovation in place.

In ensuring that the policy regimes are well coordinated at the level of conceptualization, implementation and practice, the following questions are of relevance:

(i)   How does innovation policy fit into the broader context of industrial development strategies of countries in practice?

(ii)   What are the most critical areas of coordination?

(iii) What lessons can be drawn from the experiences of countries in promoting policy coordination at the macro-, meso- and micro-levels for improved firm-level performance, and can they be understood and applied to other countries?

III. FIVE PRINCIPLES CAN GUIDE THE WAY

(i)   This report identifies five broad alignment issues that play a causative role in the overlaps, namely:

(ii)   The existing gaps in policy articulation and design;

(iii) A lack of policy coherence and policy competence in the implementation process;

(iv)   The prevalence of competition between ministries, agencies and duplication of efforts, which result in resource constraints;

(v)   Insufficient capacity to conduct policy evaluation and monitoring; and

(vi)   A lack of coordination between policymaking, governmental interventions and business environment.

It proposes five principles as guidelines to countries to find the right balance between policy processes and policy coordination. These principles are aimed at:

(i)   Identifying and eliminating policy redundancies in the policy conceptualization and policymaking structure;

(ii)   Promoting policy coherence and policy competence;

(iii) Using resources carefully;

(iv)   Developing capacity for proper policy evaluation and monitoring; and

(v)   Coordinating the policymaking processes closely vis-à-vis their impact on the business and enterprise environment, and promoting private sector engagement.

IV. COUNTRY FINDINGS REINFORCE THE IMPORTANCE OF GETTING THE POLICY INTERFACE RIGHT

In the three African countries that are the focus of this report, industrial and STI policy issues were examined against the following questions:

(i)   What are the historical, economic and systemic factors that contribute to the way STI and industrial development policies evolve in countries over time (policy conceptualization and policy history)?

(ii)   How do these historical, economic and systemic factors impact on the way policies and institutional support are structured in practice (policy coordination and implementation)?

(iii) How does this impact firm-level performance in countries (policy impact on firms and sectors)?

The country studies are detailed investigations that show how the institutionalized patterns of policy conceptualization and policy implementation (in terms of coordinating the various components of industrial development, and aligning the instruments and mechanisms to local requirements) are critical to ensure firm-level performance.

1. Factors for country selection

The country selection was based on three sets of parameters:

(i) The developmental and institutional circumstance represented by the country: While Nigeria is a commodity-rich developing country; Ethiopia is a least developed country (LDC) with a resource-concentration in agriculture. This is juxtaposed with the experience of the United Republic of Tanzania, which is a mix of resource-based activities and other sectors. As a result, each of these countries serves to illustrate a developmental challenge in the realm of coordination of industrial and innovation policies for developmental outcomes.

(ii) The ongoing policy transformation in industrial and innovation policies: All the three countries discussed in this report have national vision documents, new industrial development strategies and STI policies that embody the aspiration of its leaders and policymakers to transform their nation into 'middle-income' economies within the next two to three decades.

(iii) Difficulties faced in channeling R&D expenditure and GDP growth rates towards technological learning: All three countries have experienced relatively impressive GDP growth rates over the past decade if not longer, and increased R&D expenditure as a percentage of GDP in the 2000s. Despite this, they have faced difficulties in focusing these investments into greater technological learning, particularly at the firm level, as demonstrated by the lack of greater exports of medium and higher technology products.

2. A summary of country findings: Nigeria

Nigeria aspires to have a mature economy with a diversified industrial base, and to reduce reliance on oil-based exports, which currently account for over 90 per cent of its export earnings. Industry, the second largest sector in Nigeria, accounted for about 26 per cent of GDP in 2013, but most of this was attributable to the oil sector: out of $100 billion worth of merchandise goods exports in 2013, fuels accounted for $94 billion. The reliance of the economy on crude oil exports, which accounted for about 70 per cent of total exports during the past four decades, led to a shift away from industrial activities of a productive nature, leading to low structural change, low dynamism and over-dependence on a single commodity. Key general, sectoral and firm-level findings based on the empirical survey of 200 firms across three sectors (agro-processing, ICTs and health and pharmaceuticals), field interviews and a historical review of the country's economic development are summarized below.

a. Tracing policy conceptualization and policy history from 1960s until the present day

An in-depth policy analysis shows that the failings of development plans since the 1960s inhibited the adoption of a comprehensive approach integrating technology acquisition and training to industry. As a result of this, flailing industrial productivity led to the gradual ineffectiveness of a large number of public sector enterprises and local firms. The S&T policy adopted in 1986 and which was revised in 1997 and 2003 did not succeed in reversing the shortcomings of the national innovation system because technology was largely conceived in terms of generic acquisition of hardware machinery and equipment, rather than as a process of building technological absorption capacity. To address this, Nigeria enacted the National Industrial Policy of 1998 and simultaneously embarked upon a system-wide review of its S&T framework in 2005 to shift the focus to building innovation capacity. As a result of the review process, a new STI policy framework was launched in 2011 to harness, develop and utilize STI to build a large, strong, diversified, sustainable and competitive economy that guarantees a high standard of living and quality of life to its citizens.

Along with the 1998 National Industrial Policy, Nigeria is also guided by the Nigeria Vision 2020, which is currently being implemented through the National Implementation Plans. Nigeria Vision 2020 is a long-term strategy aimed at transforming the Nigerian economy into one of the top 20 economies by expanding the country's economy from $173 billion in 2009 to $900 billion by 2020 with a per capita income of $4,000. The review finds that past efforts in promoting industrial development in Nigeria failed largely due to a lack of focus on technological learning at the plant, sectoral and industry level. Current policy efforts seek to address this and integrate these concerns, which is a very positive development.

b. Assessing challenges for policy coordination
    and implementation

However, despite the recognition that industrial policy
and STI policy are com-plementary, survey results from
the three sectors show that firms continue to encounter
difficulties that affect their ability to perform; these ongoing difficulties stem from policy coordination and implementation issues.

This can be attributed to two issues. Both the new STI
policy and the Nigerian industrial development strategy
and implementation plans are largely being implemented
within an institutional setting in which industrial development and innovation capacity are considered as two con-
trasting goals. Furthermore, several older policy directives
aimed at changing underlying policy processes to promote collaboration and com-munication among the various actors in the institutional support system have yet to
be considered. For example, there is an indication in the
new STI policy that the National Science and Technology
Act, CAP 276 of 1977 and the Federal Ministry of Science
and Technology Act No 1, 1980 would be reviewed, but
this review had not been carried out at the time of the
survey. The mandate of the National Office for Technology
Acquisition and Promotion, which was created in 1979,
also needs to be reviewed and given a mandate to ensure
better coordination and impact.

A second issue is that both policy frameworks, despite their aims, have not yet addressed basic issues
of capacity building and infrastructure. That is, they still
remain largely concerned with articulating objectives
rather than addressing grass roots challenges. A lack of
investment into public utility services continues to hinder
the provision of good physical infrastructure for industrial
activities. Particularly, the lack of electricity and transport
infrastructure has been a hindrance to industrial production since the 1970s, when the issue of power supply
was not well-integrated into the construction of large-
scale industrial plants.

c. Measuring policy impact at the firm level

The survey results show that despite the efforts to enact
the two policy frameworks, there is not much real impact up until now on the way firms innovate, learn and
compete. The focus of their activities is in marketing and
distribution of products rather than innovative activities
that can help create new products and processes. The
survey also shows that Nigerian firms are engaged in
incremental learning activities, and often ranked their
products and processes as new to the local market, and
not to the region or the world.

Many of the firms interviewed were often unaware of the
national STI policy, or the incentives contained therein.
Companies were also unaware of new agencies that
were recently set up to assist them to compete, such as
the National Competitiveness Council. The survey also
showed that there was a low awareness of the kinds
of incentives that were available to promote firm-level
innovation, learning and competitiveness. Firms also
reported difficulties in benefitting from these schemes,
where available, due to the extensive bureaucratic processes involved.

3. A summary of country Findings: United Republic of Tanzania

The United Republic of Tanzania has recently emerged
as one of the best performing economies in Africa. This
is in marked contrast to the 1970s when the real per
capita GDP growth rate was only 0.5 per cent and which
further plummeted into negative growth rates (-0.7 per
cent) in the 1980s. However, in the past two decades,
the country's economy experienced a steady rise with
real per capita GDP growth rates, which surged from
0.9 per cent in the 1990s to 4 per cent in 2000s and 4.1
percent in 2010-2014.

Despite these trends in overall growth pattern, industry has
contributed the least to GDP growth, lagging behind services and agriculture since the 1980s. By way of contrast,
the services sector accounted for the largest share of GDP
in 2013, with a contribution of 47.3 per cent; the agriculture
and industry sectors accounted for 31.7 and 21 per cent of
GDP, respectively. The challenge therefore remains one of
fostering industrialization through technological change and
innovation. Relevant findings are summarized below based
on a three sector survey (agro-processing, ICTs and health
care and pharmaceuticals) of 144 firms, and analysis of the
policy regimes on industrial policy and STI since the 1960s.

a. Tracing policy conceptualization and policy history
    from 1960s until the present day

The 1967 Arusha Declaration served as a beacon of
policy focus in the immediate post-independence period, with implications for early industrial development
policies focusing primarily on state-led industrialization
through local, indigenous efforts. However, by the end
of the 1970s, failures to boost industrial capacity were
attributed to a low focus on technological capacity. This
not only led to the establishment of the Tanzania Commission for Science and Technology in 1986, but also
the national S&T policy that was formulated in 1996.

However, the 1996 S&T policy suffered from certain
shortcomings, the most im-portant of which was insuf-

ficient focus on technological learning and innovation. Sectoral objectives and strategies were also not fully translated into policy actions and investments in knowledge infrastructure were not realized as intended. This led to a continued disconnect between industrial and innovation policy frameworks in the country.

Additionally, since the 1980s, the United Republic of Tanzania also underwent a few re-orientations of its industrial policy. The earlier import substitution policies were replaced with a market-oriented approach in the late 1980s, along with trade liberalization of the economy. Trade liberalization resulted in a large-scale exit of local firms from the Tanzanian market due to a lack of institutional support for industry and their inability to compete with foreign firms. In an effort to revive the local industrial sector, the government sought to promote an industrial strategy focusing on high-technology sectors, as in the East Asian economies. Lacking donor-support, this plan was replaced with a National Strategy for Growth and Poverty Reduction (NSGRP 2005-2010), which focused primarily on poverty reduction. An integrated industrial development strategy was also enacted since 2011, along with the National Development Vision 2025. Currently, the United Republic of Tanzania is in the process of implementing its second five-year plan to further these objectives.

In order to achieve the targets set out in the industrial development strategy, a revised national STI framework was tabled in 2013, and is pending approval of the Cabinet.

b. Assessing policy coordination and implementation

Despite recent efforts to consolidate industrial performance, there is a lot of policy incoherence in the design and articulation of policies on the one hand, as well as the implementation of policy mandates on the other. A lack of connectedness among the industrial development plans, sectoral strategies and the national S&T policy, coupled with the absence of a plan to guide the coordination of these policies, continue to hinder the country's development. There seems to be an urgent need to implement the new STI Act, and also to coordinate industrial development with technological change and technology transfer. This is currently being considered a priority by the national planning commission for the second five-year plan (set to be enacted sometime in 2016).

The survey and interviews showed that the coordination shortcoming related to the roll-out of these plans, strategies and policies are in large part similar to what was ob-

served in the 1990s between the S&T policy, industrial policy, finance, education, etc. As a result, although the policy imperative is to boost local production capacity or expand the industrial base, this is compromised by a lack of institutional coordination. Meanwhile, despite the new integrated industrial development policy of 2011, a shortage of emphasis on technological learning, low absorptive capacity and low emphasis on innovation continue to hinder industrial development, particularly in the manufacturing sector.

These shortcomings have, to a large extent, negatively impacted industry. At the sectoral level, manufacturing activities went into a steady decline since the 1990s and accounted for 7.2 per cent of GDP in 2013, with the bulk of industrial growth being accounted for by non-manufacturing sectors, such as mining and construction. The manufacturing sector was characterized by the creation of low-value added products for the domestic market and export-oriented activities with little or no productivity growth.

c. Measuring policy impact at the firm level

Over 88 per cent of industry is comprised of micro-enterprises with less than five workers, which contributed a third of the country's GDP. Overall, most of the industry is made up of informal, micro- and small-sized firms, with a few medium and large-sized companies. Further, the majority of the micro- and small-sized medium firms operate in the services sector, while the rest are in agriculture and manufacturing.

The survey found that at the firm level, few businesses were engaged in innovation activities. Most of the small-scale firms were engaged in in-house operations relying on local and often self-sourced financing. Lack of finance, in particular, has prevented firms from undertaking technological development and innovation. Also, firms focus on short-term activities on how to survive and sell their products because of the uncertain innovation and industrial environment in which they operate and lack of support impedes their ability to innovate.

Survey data showed that a lack of policy coherence on various aspects of industrial and STI policies, such as levies imposed on imports of raw materials (as opposed to an exemption of levies on final products) in some sectors served as a disincentive to innovate or manufacture locally.

In addition, firms reported receiving little in the way of government support to participate in innovation and finance schemes. Firms also found that regulatory frameworks were often very hard to navigate, and that this contributed to a large informal sector characterized by low technological capability and lack of investment in R&D. Finally, shortcomings in the innovation environment affected firms to a large extent. Currently, firms have little or no interactions with universities, public and private research institutes and other intermediate organizations. This hinders technological learning in both the public and in the private sector.

4. A summary of country findings: Ethiopia

Ethiopia has recorded impressive economic growth over the past two and half decades. The real per capita GDP growth rate rose from -1.4 per cent in the 1980s to 2.3 per cent in the 1990s, peaking at 6.7 per cent between 2010 and 2014. Ethiopia's current challenge remains one of diversifying its economic base, and strengthening its economic performance. The bulk of the Ethiopia's GDP value added has come from the primary sector comprising agriculture, hunting, forestry and fishing, which jointly accounted for 45.5 per cent of the GDP value added in 2013. At the sectoral level, the key challenge is one of increasing the share of GDP value added from industry, which has not only been less than agriculture and services over time but its share of contribution has also declined in the past four decades from 16.2 per cent in 1973 to 11.1 per cent in 2013.

General findings, as well as sectoral and firm-level findings, are summarized below based on a survey of two sectors (agro-processing and pharmaceuticals) and a historical review of the industrial and innovation policy frameworks.

a. Tracing policy conceptualization and policy history from 1960s until the present day

Detailed policy analysis shows that Ethiopia's recent economic success has been shaped by the country's developmental plans over the past two decades, the most relevant of which is the Growth and Transformation Plan (GTP). This five-year economic master plan was launched in 2010 and aimed at achieving 11-15 per cent annual GDP growth and large-scale investments in industrial and agricultural sectors by 2015. A second phase of the GTP, the GTP II, is due to be launched in 2016 to cement and build on current achievements.

Along with the GTP 2010-2015, Ethiopia also sought to revive and resuscitate Ethiopia's S&T policy framework. The STI framework was fragmented since its creation, which despite the formulation of the first national S&T policy of 1993, and the re-establishment in 1994 of the Ethiopian Science and Technology Commission as an autonomous public institution was not entirely

addressed. A fundamental weakness of the 1993 S&T policy (which was later amended in 2006 and 2010) was that it was narrowly focused on S&T without any emphasis on innovation capacity. Furthermore, the policy envisaged no coordination with industrial development at the sectoral and plant levels. A revised policy of 2012 now seeks to focus attention on innovation and technology transfer, in conjunction with the creation of a centralized innovation fund for R&D activities, which was established with the aim of committing at least 1.5 per cent of the GDP annually to applied research.

The GTP 2010-2015 and the STI policy are well coordinated in their goals, and the GTP reinforces the issue of building capacity in the local context by placing emphasis on the development of universities, research institutes, technical and vocational education and training institutions. Programmes have been defined that promote these linkages namely: (a) the development of industrial zones; (b) capacity building programmes; (c) university-industry linkages; and (d) the creation of a centralized R&D and innovation fund.

b. Assessing policy coordination and implementation

The share of investment in manufacturing activities has been impressive, wherein Ethiopia approved 1,211 projects for the manufacturing sector in 2011/12, which accounted for 31 per cent of the share of total investment capital over this period. The central challenge now is to ensure policy coherence and coordination between industrial and innovation policies at the implementation level, which still remains weak. Particularly, there needs to be a greater emphasis on the provision of a common STI infrastructure, technology-transfer venues and information sharing of relevance to promote the industry, especially to engage in high technological intensity activities.

Policy coordination and implementation is still less than satisfactory because the institutional apparatus in the country remains weak and fragmented in this regard. The survey and analysis found that a large number of intermediary agencies such as those that can help industry acquire and upgrade technologically are missing, or just being set up. A good case is that of the Food and Beverages and Pharmaceuticals Industry Development Institute, which has recently been set up to promote such linkages recently.

c. Measuring policy impact at the firm level

The limitations of policy coordination and implementation are felt at the firm level, as the survey findings show. The results show that at the firm level, there is a lot of capacity in Ethiopia's agro-processing activities beyond coffee production, e.g. several firms are engaged in leather activities, but these activities are dominated by SMEs. The survey also found that firms face significant difficulties in diversifying into technology-intensive activities, especially those that can contribute to value-additions.

The difficulties faced by firms are partly due to a lack of adequate institutional support to develop technology and innovation capacity as a whole. As a result, most companies (even those in the agro-processing sector) continue to focus on domestic market opportunities, and only a few have ventured into markets beyond Ethiopia. The survey also found that firms rely heavily on not so up-to-date equipment and machinery, but some are acquiring new knowledge through the acquisition of new machinery and equipment, even though the lack of technological absorptive capacity hinders their ability to innovate. Promoting technology transfer, access to finance, joint ventures for production and value-addition remain really important to firms.

V. WHAT MATTERS IN PRACTICE: FINDINGS AND RECOMMENDATIONS

The difficulties in coordinating policy objectives, implementation and impact, as faced by the three countries in the report, are not isolated issues. A large number of countries in the developing world are faced by the same kinds of issues. Some general findings stand out in this regard. Firstly, although there have been laudable efforts in defining policies, simple infrastructure issues that have impeded industrial development over a period of decades have not been resolved. This should be the first area of focus. Secondly, countries continue to face difficulties in coordinating implementation – a development that can be traced back to the lack of policy coherence. This is not to say that ministries and agencies have not been well intentioned. In fact, the survey found that despite their best intentions and efforts, firms were not benefiting from these efforts due to a lack of policy coordination. This reinforces the need to get the policy processes right. Other more specific results on the interface of industrial-innovation policy are presented below, with accompanying recommendations.

1. There are several gaps in the policymaking structure

In all three countries, as is the case with a large number of other African countries that are also reviewed in the report, national STI policies either evolved much later (at least two decades after the industrial development policies were enacted), or evolved in parallel with little or no coordination with established industrial development frameworks.

The report finds that within countries, a predominant issue is where industrial policy is placed, and how it is articulated. In the case of a large number of developing countries, policies for industrial development are not usually articulated as industrial policies, but rather as industrial development strategies, or as national visions, or as part of recurring national developmental plans aimed at facilitating overall development and economic transition.

If countries enact national visions that include industrial policy objectives (which is the case not only in Ethiopia, Nigeria and the United Republic of Tanzania, but also true for a large number of other African countries), it needs to be borne in mind that such national vision statements generally have a broader scope than just promoting industry, and often tackle issues of poverty, youth, environment, employment and urbanization. In several countries, industrial development objectives are embedded in their national development plans, and are often recurrent on a term-by-term basis.

Therefore, although such visions or strategies encapsulate the main industrial objectives or goals, there is a need to have clear roadmaps to achieve these visions, with accompanying targets, so that these can be linked to a policy implementation mechanism on the one hand, and to STI and other policies (covering areas such as trade, investment, and development) on the other.

Another reason for the gaps in policymaking is that a large number of industrial development strategies are one-dimensional: they target overall industrial development and an increase in per capita GDP growth rates, or a rise of specific sectors. The focus should instead be on closing the productivity gap, i.e. how to ensure greater returns from productive activities. This leads to gaps in policymaking, including a neglect of:

- Technological and technical support systems required for the growth of sectors;

- Links between the human skills requirements of the various sectors with enhanced performance projections;

- A clear articulation of how the higher GDP spending on R&D will form part of public sector assistance to technological upgrading, e.g. the establishment of common industry services, technological incubation, industrial research labs, etc.

2. Policies suffer from inconsistencies and often, overall incoherence

A key issue that stands out is that sophisticated policies are not sufficient. While industrial development strategies in the selected countries recognize the importance of technology-led growth, and whereas all STI frameworks recognize the importance of coordinating with industrial policy, the same historical patterns of lack of coordination between innovation and industrial policy frameworks persist. Countries have tried to tackle these issues by providing for common goals or missions in the two policy frameworks, but policy incoherence often occurs at the stage of policy articulation, and is also often deeply rooted in policy implementation processes.

The country chapters help to illustrate the main finding of the analytical framework, namely that it is crucial that policy *processes* are clearly laid out. Specifically, the findings show that even elaborate policy frameworks on STI policy and industrial development need to be accompanied by policy consistency and coherence at the levels of:

(a) Policy conceptualization and design;

(b) Policy implementation and coordination

A number of reasons explain the existence of policy incoherence and inconsistencies. The country chapters show that they could be the result of ineffective policy transitions (where countries embark on changes in policy, but remain incomplete and lose momentum as a result of changing political leadership at different levels of governance), institutional inertia and resistance, or a lack of policy competence to foresee and avoid overlaps. A second form of policy incoherence is when the frameworks are overarching but not accompanied by a concrete implementation plan. However, in many other cases, policy frameworks are accompanied by implementation mechanisms, but several shortcomings have prevented them (to a different extent in the three countries) from achieving an impact. A key issue (already raised in the previous point) is that in the absence of stocktaking and attempts to streamline the institutional apparatus, many public sector agencies have mandates to implement the policies. When the policy framework is not completely consistent or accompanied by clear implementation mechanisms, the country analyses show that there is no clarity at the policy implementation stage as to which of the existing agencies should implement the mandates contained in the policy framework and how they should be implemented.

a. Policy incoherence in conceptualization can be a result of ineffective or slow policy transitions

Moving towards an innovation policy is a challenging coordination task, and not just one of providing a regulatory framework. In reality, although a wide variety of policies emphasize 'innovation', field investigations show that while some policies seek to fundamentally chart new ground, in some other instances, the policies often make reference to 'innovation' but are not comprehensive enough to tackle the difficulties of fostering innovation. Furthermore, there are difficulties imposed by the fact that policy processes are not followed through, and maintained during and after political transitions in countries.

The same difficulty holds true for industrial development policies. Sudden policy shifts that do not promote a coherent notion of industrialization as a continuous process lead to policy inconsistency and incoherence simply because they do not offer a consistent and reliable level of support to the process of industry transformation.

b. Policy incoherence can be due to institutional resistance and inertia

The field interviews and surveys shed light on the fact that policy and institutional history matters. Historical analyses of the evolution of policies and implementation mechanisms conducted in the chapters shows that agencies implementing these mandates operate within weak, unaccountable implementation processes. Such inter-agency rivalries exacerbate policy coordination issues and have led to a large-scale neglect of the private sector. In almost all countries surveyed, private sector enterprises considered that existing policy frameworks and the actions of implementing agencies operated at a distance from them, making little attempt to liaise and understand the constraints they faced or tried to alleviate them. Such institutionally embedded habits and practices often offer severe resistance to newer more collaborative modes of interaction. Policies on industrial development, if they are to be coherent with innovation policies, should seek to address the operative mandates of agencies to promote a change in mindset.

c. Policy incoherence can be due to insufficient policy competence / policy foresight

Another set of coordination issues arise from the fact that both industrial development and innovation policies often identified targets and objectives that were impacted upon by other policies differently. For example, in Ethiopia, the STI policy aims to 'develop, promote and commercialize useful indigenous knowledge and technologies'. To promote this, there would normally be a need to assess whether the *sui generis* system created by the Ethiopian 2006 Proclamation on Access to Genetic Resources and Community Knowledge, and Community Rights could help protect useful indigenous knowledge and technologies. In other words, the IPR protection has to be integral part of the indigenous knowledge commercialization process. But what appears to be missing in the objectives are strategies to create STI policy awareness at all levels of government, including the Cabinet and Parliament, as well as to build an innovation culture among businesses, the youth and society at large. Similarly, one of the projects under the GTP is the establishment of industrial parks, but these are expected to act as hubs for FDI, and to leverage technology transfer of the kind outlined in the country's STI policy. This once again calls for coordination of policy implementation on a strategic basis between the ministries, as well as agencies implementing the mandates on industrial development, investment and STI. But often the lack of policy competence, as well as a lack of incentives on part of the agency employees leads to very minimalistic interpretations of these mandates.

d. Recommendations to improve policy coherence in conceptualization and design

Assessing the successes and difficulties faced by the countries in this report, the following recommendations are suggested to avoid this kind of policy incoherence:

- Policy vision, mission and objectives should be closely aligned: The review of ongoing initiatives at the African level, as well as the country chapters lend strength to the conclusion that a close alignment of industrial development and innovation policies is still an elusive goal in countries. Oftentimes, even the targets or objectives for STI mentioned in industrial policy are not the same as the objectives of the STI policy itself (see previous point), thereby promoting policy incoherence and leading to confusion.

- Emphasis should be placed on developing local linkages and unlocking learning potential: Although STI policies clearly lay down the broader vision to build capacity, fostering an innovation ecosystem calls for emphasis on the creation of an innovation and entrepreneurship culture with concrete links to industrial development. It is necessary to promote entrepreneurial programmes, align academic curriculum with entrepreneurial needs, and introduce entrepreneurship classes at schools and institutions of higher learning to enable the effective application of new technologies and innovation for industrial development. The GTP in Ethiopia, for instance, has at least two such projects on building capacity.

- While enacting new policies, there is a need to clearly link them with existing initiatives and agency mandates: The country chapters found that although national policymakers are aware of the need to review existing policies and agency mandates, change is usually slow, leading to policy ineffectiveness, as in the case of Nigeria. Making this happen alongside the policymaking/revision process is critical for at least for two reasons: Firstly, previous policies often have agency mandates that call for review in the light of the new policy, to ensure that the institutional framework embodies the changes in a dynamic and efficient way. Secondly, reviewing policy mandates is very important to ensure that national resources, particularly financial resources and human skills, are used efficiently.

e. Recommendations to improve policy coherence in in the implementation process

The recommendations in this regard include:

- Coordination hurdles need to be tackled at the level of agencies and organizational structures in order to avoid overlapping mandates between newly created agencies and existing agencies, and how they interact with the private sector. Duplicated measures should be taken stock of, and efforts should be made to eliminate such duplication over time.

- Policy changes should be accompanied by appropriately funded and transparent budgets and staffing of skilled employees to facilitate their implementation.

- Schedules and critical milestones to be achieved jointly by the STI and industrial policies should be clearly defined ahead of the process, and also framed in a manner that addresses national needs and industry characteristics.

- A high-level governance structure and coordination matters, especially at the ministerial level. More efforts should be made to ensure such interaction.

- Best practices from other countries can only serve as a guideline; the right combination of innovation and industrial policies is a personal choice of countries.

- The focus should be on contextualization in order to achieve results.

3. Policy monitoring and evaluation mechanisms are required to ensure efficient use of existing resources

Monitoring and evaluation (M&E) mechanisms are relevant from a variety of per-spectives. They not only enhance coordination efforts but also point to the lack of funding of various initiatives as part of the stocktaking process. They also ensure that funding issues are taken into consideration and reviewed over time to evaluate: (a) where is the current funding being used? (b) What are the funding gaps to implement the goals of industrial and STI policies? (c) How can the gap be financed? (d) What are the best ways to share risk and partner with industry to effect transformation? (e) How to best allocate existing resources, and into what agencies? (f) Can agencies be streamlined and better defined? These are some of the issues that should form a core part of the monitoring and evaluation exercise.

Monitoring and evaluation exercises aimed at ensuring that existing resources and agency strengths are put to good use will play a pivotal role in policy effectiveness.

In support of this point, the surveys and interviews showed that most funding given to agencies supporting innovation is often spent on recurring expenses related to staff maintenance and running costs, with little or no reserve for innovation support infrastructure. In the United Republic of Tanzania, for example, about 95.1 per cent of the sums allocated to agricultural R&D goes into staff salaries or operating expenses, leaving only 4.9 per cent for capital investments in 2011. Similarly, staff salaries and operating expenses account for about 83.4 per cent and 71.8 per cent of agricultural R&D in Nigeria and Ethiopia, respectively.[1] Similarly, supporting staff account for about 29.3 per cent (2010), 33.6 per cent (2007) and 37.9 per cent (2010) of the R&D expenditure in the United Republic of Tanzania, Nigeria and Ethiopia, respectively. By way of comparison, the share of support staff in relation to R&D personnel is smaller in other developed countries, e.g. Germany (16.8 per cent in 2011) and Japan (16.2 per cent in 2011), as well as in other developing countries with highly sophisticated R&D system, e.g. Hong Kong, China (5.5 per cent in 2010).[2]

a. Recommendations to ensure efficient use of existing resources

In order to address these issues, the following recommendations could be con-sidered:

- There is a need to integrate monitoring and evaluation from the start of the policy process.

- There is a need to ensure monitoring and regular follow-up, along with open assessments of budgets and assistance offered by various agencies.

- Monitoring and evaluation should be based on institutional memory of why and how coordination failed, because looking inwards to assess and apply the learning of the country's own past as to why policies failed or what factors vitiated the policy processes helps to promote successful coordination.

- The resources earmarked to support the implementation of relevant policies will largely determine the effectiveness of the policy in question. Hence, policies should be accompanied by resource allocations that are on par with the activities envisaged.

4. Policymaking, government interventions and the business environment should be coordinated more closely

An important finding of this report is that policy is often reality-incoherent. That is, as opposed to the practical structure of the local industry, which is often overwhelmingly comprised of SMEs and the informal sector, industrial policy and innovation policy elaborate sectors of importance that are entirely high-tech, or require an institutional infrastructure that is very far-fetched from the on-the-ground realities that firms face in their day-to-day existence. A number of the local firms are operating on the fringes of technological development even in the so-called high technology sectors. For example, in the ICT sector, many companies simply offer call management or ICT services to users (as opposed to any production or process improvements), in the pharmaceutical sectors, many companies only distribute already packaged medicines, or engage in traditional medicine-based preparations of low-technological nature.

It is important to bring the private sector into the policy focus and the realm of policy discourse in the countries. The STI and industry policy frameworks should be adequately accompanied by both business and industry support organizations, which provide incentives for local firms such as R&D grants, R&D loans, tax credits and governmental procurement, all of which have met with much success in other developing countries. In fact, one of the key issues that were raised in the country studies related to the way the question of finance was addressed.

Countries, such as Thailand, have used policy mechanisms like government procurement as an incentive for innovation.[3] Incentives such as these could be considered in all the three countries there were policy implementation gaps on the question of innovation finance.

* * *

African countries are at a defining point of stocktaking, particularly as they transition into an era of new development goals. It is becoming widely acknowledged that sustainable development rests more broadly on stable industrial development of a kind that can deliver better livelihoods to the people and eradicate poverty, as several goals of the recently adopted 2030 Agenda for Sustainable Development emphasize. In particular, Goal 9 encapsulates the dual objectives of promoting inclusive and sustainable industrialization and fostering innovation.

Almost all countries in the African region, and more widely in the developing world, including the three countries that were studied in depth for this report, are currently at a policy and developmental stage where industrial development through technological change *should be* a central, if not the most important, priority. Not only is there a policy transition towards that end, the field surveys were testimonies to the extent of political commitment to enacting elaborate industrial policy frameworks, and revising their S&T policies towards policies dedicated to innovation. But the private sector in the African region (particularly in sub-Saharan Africa) is in dire need of greater support, and enterprise policies are currently the weak link.

NOTES

1. ASTI website (http://www.asti.cgiar.org/countries) accessed on 27 April 2015.

2. UNESCO Institute for Statistics database (http://data.uis.unesco.org/) accessed on 27 April 2015. Full time equivalent (FTE) figures were used.

3. See UNCTAD, Promoting Innovation Policies for Industrial Development in Thailand, Forthcoming.

INDUSTRIAL DEVELOPMENT AND INNOVATION POLICY

1

CHAPTER I
INDUSTRIAL DEVELOPMENT
AND INNOVATION POLICY

A. INTRODUCTION

Industrial development or industrialization is a process whereby labour and resources gradually shift from agriculture to manufacturing, leading to a steady rise in productivity rents and overall economic development. A key purpose of industrial policy is to promote this process of structural transformation by targeting economic activities, sectors and technologies with growth and development potential.[1] It has come to be accepted in countries as both an instrument of growth and transformation, and as a lever to promote innovation capacity and inclusive social prosperity.

In practice, industrial policy is an umbrella framework of a rather broad nature. It can comprise interventions that impact industrial development, but also include policies affecting science, technology and innovation (STI), foreign direct investment (FDI), intellectual property rights (IPR) and trade (Cimoli, Dosi and Stiglitz, 2009). Given this wide scope, there are potential areas of relative overlap between industrial policy and several other national policies. One of the largest overlaps of this nature is between industrial development and STI policies, as the former aim to promote a 'great transformation' by facilitating capabilities for knowledge accumulation within firms and sectors. As a result, the incentives and instruments of both policies are often quite similar and aimed at facilitating technological learning and innovation.

Although industrial policies are not entirely new to developing countries, such policy overlaps matter; making it critical to resolve these overlaps in order to align industrial and innovation policies in such a way that they are mutually supportive. There are several reasons why this is important. The first reason can be traced back to the justification of industrial policy itself. Industrial policies seek to catalyse institutional change by addressing existing shortcomings in information or coordination. But when implemented in conjunction with other contravening incentives or policy interventions, it may not only fail to achieve this objective, but may often even lead to misuse of scarce resources, hence focusing attention on better coordination.

The second reason is that the accumulation of capabilities for knowledge creation relies on skills and technical abilities, which is not an easy process. There is a range of other supportive institutional infrastructure that play a role in the way skills and technical competence are used to create new products and processes by local firms. Such supportive infrastructure is critical in enabling the development of capabilities through linkages between various actors, thus shaping how firms respond to learning opportunities, benefit from collaborations, and enhance technical efficiency of production. No unique set of policy prescriptions exist that can be shared, but the experiences of some East Asian countries or industrialized countries offer some clues on how this can be achieved. A key lesson in this regard is that complementary, reinforced incentives between industrial and innovation policy frameworks are very effective in promoting industrial upgrading (see Aiginger, 2014; Amsden, 2001).

The third reason is that while there are some good examples of developing countries that have historically coordinated their industrial development strategies with STI policy objectives, there are also an equal number that appear not have managed to do so. They have approached industrialization and technological change as two different elements in the developmental process. In practice, this has led to a dual narrative on industrial development strategies and STI policies.

This report aims to approach the linkages between innovation and industrial policies from a practical perspective in order to highlight the pitfalls and the opportunities for developing countries. The 2030 Agenda for Sustainable Development aims to achieve sustainable and inclusive industrial development; to this end, eliminating policy overlaps will be key to achieving these goals. Fundamentally, at a broader level, it seems almost intuitive to assume that STI/ innovation policies are an integral complement to industrial policies/ national industrial development strategies, and achieving sustainable industrialization calls for both frameworks to be coordinated very closely. At the same time, coordination is not an isolated 'policy-related' goal, its impact is felt by, and a critical element in, the performance of all firms, as the aggregate productivity of an industrial sector is the sum total of the productivity of the individual firms represented in this sector.

A synergistic environment for innovation-led industrial development rests on coordination of policy implementation at three different levels, macro-, meso- and micro. At the macro level, i.e. at the level of national foresight and policymaking, it should be noted that policy frameworks on both industrial development and STI policy should be articulated to provide a lean and cogent conceptualization of common goals and objectives. The coordinated implementation of these policy frameworks occurs at meso-levels, i.e. when the policies are converted into implementation through incentives, programmes and agency mandates. Their impact, however, on firm-level performance is often an issue that occurs at the ground/ grassroots level, i.e. it is a micro-issue that is affected by a range of factors that impact day-to-day performance. Even if relevant policy frameworks on industrial development and innovation are in place, without coordination at all three levels, they may not be well coordinated at meso-levels, thereby negatively impacting on firm-level performance and vitiating the common goal of promoting technology-led industrial growth.

Institutional environments and factors that vitiate coordinated policymaking are often path dependent: they depend on the manner in which countries evolve, implying that what is observable today may draw upon history, and that such factors are 'embedded' in the underlying systemic context of the economy (Evans, 1995). The ways in which policies interact and how systems engage in problem solving are often shaped by historical, cultural and social parameters. This report therefore seeks to bring new light on how historical factors shape coordination between industrial development and innovation policies within countries.

The following key questions will be considered in the course of this report:

(i) What are the historical, economic and systemic factors that contribute to the way STI and industrial development policies evolve in countries over time (policy conceptualization and policy history)?

(ii) How do these historical, economic and systemic factors impact on the way policies and institutional support are structured in practice (policy implementation)?

(iii) How does this impact firm-level performance in countries (policy impact on firms and sectors)?

In order to thoroughly investigate these questions, the report studies the experiences of three African countries in promoting learning, knowledge accumulation and industrial development. The country studies presented in this report bring to light the extent to which institutionalized patterns of policy conceptualization and policy implementation (in terms of coordinating the various components of industrial development, and aligning the instruments and mechanisms to local requirements) are critical to ensure firm-level performance.

B. SCENE SETTING: THE NEED TO COORDINATE INDUSTRIAL AND INNOVATION POLICY FRAMEWORKS

The proliferation of newer industrial policies and strategies focusing on leveraging innovation need to be well coordinated with STI policy frameworks. There are a least three reasons that lend strength to such an assertion, and relate to:

(i)   A greater emphasis towards innovation and innovation rents in the global landscape;

(ii) Stagnating growth rates, or growth rates based mainly on an expansion of unproductive sectors; and

(iii) A refocus on industrial policy as an instrument to leverage change.

1. Shifting emphasis towards innovation in the global landscape

Although economists continue to face difficulties in measuring technological capacity accurately, a consensus exists that innovativeness, along with the capacity to capture related rents, are only to be found in handful of countries worldwide (Archi-

bugi and Michie, 2002; UNCTAD, 2012). Figure 1.1 helps to illustrate this clearly as it captures the role that is being played by various regions and country groupings in different product categories. For example, Asian developing countries accounted for almost half (48.9 per cent) of total global exports of high technology intensity products in 2014. This, along with the growing knowledge component of global economic activity, conveyed through terms such as the 'knowledge economy' and the 'growing technological divide', has meant that countries with little or no technological competence will inevitably face difficulties in promoting economic development.

In the current global context, the inability to promote technological change has resulted in capacity lags within countries. This manifests itself through differences in intersectoral and inter-industry productivity, and the microeconomic allocation of future resources to R&D and innovation efforts (Cimoli, Dosi and Stiglitz, 2009). Figures 1.2 and 1.3 help to show these capacity lags between different regions. The figures show the different shares of exports in medium and high-technology intensity manufactures, which are mainly attributable to the differences in knowledge accumulation capabilities.

Figure 1.1: **Distribution of world exports by technology intensity and by development status, 2000 and 2014 (in per cent)**

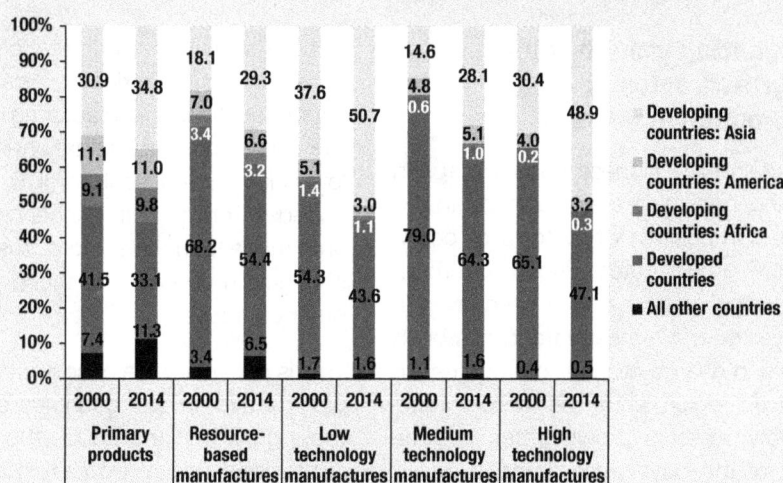

Source: UNCTAD calculations based on UNCTADstat (accessed on 15 July 2015).

Figure 1.2:　Distribution of medium-technology manufacturing exports by different country groups, 2000-2014 (in per cent)

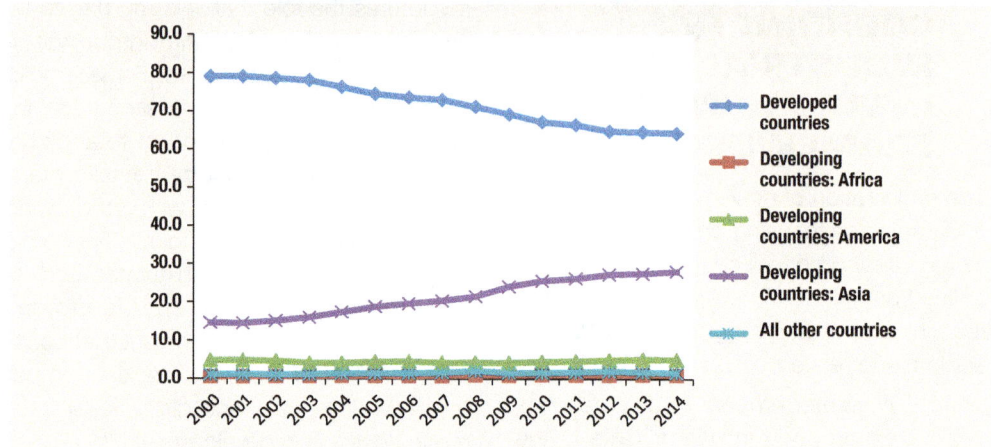

Source: UNCTADstat (accessed on 20 Oct 2015).

Figure 1.3:　Distribution of high-technology manufacturing exports by different country groups, 2000-2014 (in per cent)

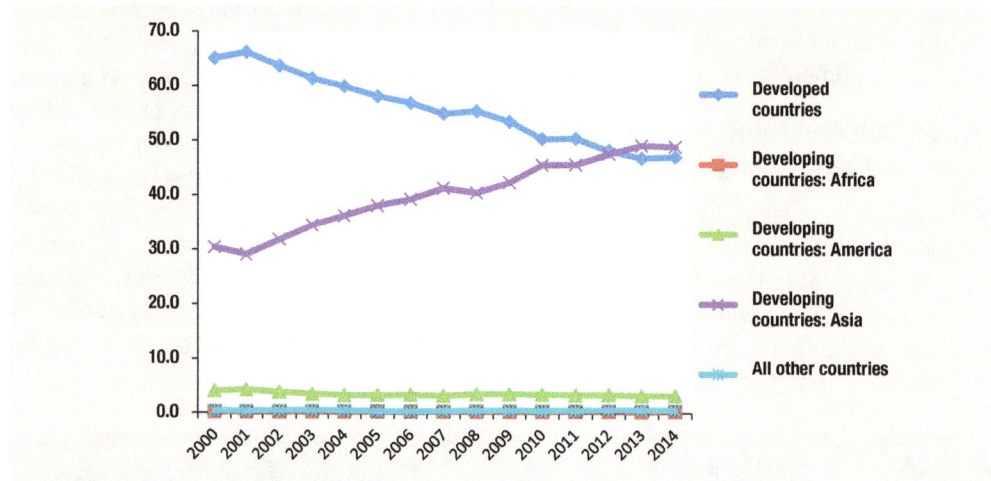

Source: UNCTADstat (accessed on 20 Oct 2015).

2. Stagnating growth rates or growth rates in unproductive sectors

To a certain extent, these capacity lags in promoting innovation are also linked to the sectoral composition of exports of countries. GDP growth rates among a number of developing countries have been rising in recent years, and a steady trend has been observed in a large number of countries in the African region since the 1990s. Table 1.1 below presents growth rates in three regions of the developing world. However, regions specializing in sectors that are productivity enhancing have experienced a more sustained rise in real growth rates, for example, countries in Asia who have expanded their manufacturing sectors. At the same time, specialization in natural or low-value added sectors (IDB, 2010) has impeded productivity enhancing growth, and made countries more susceptible to changes in demand for these products in other countries.

This is especially the case in the African region, where several countries experienced rapid growth in the 2000s, up until 2008, largely because of increasing demand for commodities or products with low-value added content (such as agro-products).

GDP growth rates in the African region were therefore more vulnerable and dependent on the expansion of economic activity in other countries, particularly in emerging economies. As a result, in the aftermath of the global economic downturn of 2007-2008, growth rates in most regions in the developing world have returned to normalcy at a faster pace than in Africa (see table 1.1 below). The slow recovery of African economies showcases the dependence of their recovery on global markets and, more recently, markets in emerging economies, which are equally important trading partners and export destinations for the African region (see UNCTAD, 2010, 2013).

Other data and existing analysis reinforce the conclusion that growth patterns that are not based on technical change or on diversification do not promote the process of continuous change along specific technological trajectories, or promote the movement into new sectors and activities that embody new technological paradigms (see Hidalgo et al, 2007).

3. Synergies between industrial and innovation policies

These two trends have forced a rethink on how countries can harness international trade to play a powerful role in reducing poverty and promoting development, through national policies that "promote a development-driven approach to trade *rather* than a trade-driven approach to development" (UNCTAD, 2004, p. 67; emphasis added).

Particularly, the growing technological divides and continuing challenges in building capabilities have led to a remarkable shift from science or S&T policies to STI policy frameworks

Table 1.1: Real GDP growth rate by region, 1980-2014 (in per cent)				
	1980-89	1990-99	2000-08	2009-14[1]
Developed countries	3.2	2.5	2.2	1.5
Developing countries	3.6	4.9	6.3	5.3
Developing countries: Africa	2.2	2.4	5.7	3.5
Developing countries: America	1.7	3.2	3.8	3.4
Developing countries: Asia	5.4	6.4	7.4	6.2
Least developed countries	2.5	3.2	7.4	4.8
Least developed countries: Asia	3.2	4.8	6.7	5.6
Least developed countries: Africa[2]	2.2	2.3	7.6	4.4

Source: UNCTADstat (accessed on 19 October 2015).
[1] 2014 figures are estimates.
[2] Includes Haiti.

Box 1.1: From S&T to STI policies

Most developing countries embarked upon S&T policies from the 1960s onwards as they were seen as appropriate instruments to build local technology capacity for industrial development. With the exception of a few countries, particularly the East Asian economies and other developing countries, e.g. China, India, Brazil, most developing countries did not integrate S&T policies within their industrial development/national development plans. This shortcoming, in combination with existing institutional weaknesses, led to a low focus on innovation.

While the predominant focus of such policies was on acquiring technologies, the lack of capabilities in terms of technical skills, creativity and innovativeness that can be created only through learning activities (through training and learning-by-doing) remained. This meant that the mastery of new technologies remained a major challenge. In addition, in order to become competitive, developing countries enterprises needed to nurture the capacity to creatively adapt and innovate, rather than merely rely on importing technologies from elsewhere.

Newer STI policy frameworks in countries seek to address these challenges through incentives for collaborative learning, networking and R&D within national innovation systems.
Source: UNCTAD.

in a large number of developing countries over the past two decades (see box 1.1).

These policies seek to place a greater emphasis on innovation, national R&D spending and knowledge expansion (see table 2.2, chapter II). For example, it was estimated that by 2010 there were up to 40 ministries overseeing various STI-related activities across various countries in Africa (UNESCO, 2010). African countries with relatively fairly advanced innovation structures include South Africa, Tunisia and Mauritius, all of whom started to develop national innovative initiatives and plans in the 1990s.

Recognizing the importance of R&D spending in boosting economic performance, the Eighth Assembly of the African Union Summit (29-30 January 2007, Addis Abeba), called for increasing R&D spending in African countries to 1 per cent of total GDP by 2010. Recent figures show that some countries have managed to increase R&D investments, but that others are not reaching this objective (see NEPAD, 2010; table 1.2).

Table 1.2:	R&D expenditure as a share of GDP in selected countries	
Country Name	**Year[(1)]**	**Share (per cent)**
Ethiopia	2010	0.25
Tanzania, United Rep.	2010	0.52
Nigeria	2007	0.22
Brazil	2011	1.21
China	2012	1.98
Ghana	2010	0.38
India	2011	0.81
Kenya	2010	0.98
Korea, Rep.	2011	4.04
Russian Federation	2012	1.12
South Africa	2010	0.76
Thailand	2009	0.25
Uganda	2010	0.56
World	2011	2.13
Sub-Saharan Africa	2007	0.58

Source: UNCTAD Calculations based on WDI Database (accessed on 7 May 2015).
[(1)] *Latest available year*

In parallel, industrial policies/ national industrial development strategies and visions and plans, envisage that economic development will be achieved by placing a greater emphasis on technological learning (see table 2.1, chapter II). Such industrial development policies/strategies also play a determining role in how the state and the private sector can collaborate to remove barriers and promote productivity growth and technological upgrading within and across sectors (see also Rodrik, 2004, Chang, 2011, Stiglitz, 2014).

Industrial and innovation policies are often synergistic in the goals they seek to achieve: technology-led industrial development. However, in practice, success in harnessing synergies between these two policy frameworks for tangible results depends less on policy emphasis and more on the *policy processes* that are put in place to ensure the accurate implementation and coordination of policy incentives. These policy processes will determine whether a set of industrial and innovation policy interventions will succeed or not (Rodrik, 2004; 2014; Aiginger, 2014; Aiginger and Böheim, 2015). In order to prevent policy conflict, or fragmentation of implementation efforts, several options have been suggested to promote this kind of policy coordination, including that countries' consider an integrated complementary industrial-innovation policy framework to ensure cohesion (see Mazzucato, 2013).

In ensuring that the policy regimes are well coordinated at the level of conceptualization, implementation and practice, the following questions are of relevance:

(i)   How does innovation policy fit into the broader context of industrial development strategies of countries in practice?

(ii)   What are the most critical areas of coordination?

(iii)   What lessons can be drawn from the experiences of countries in promoting policy coordination at the macro-, meso- and micro-levels for improved firm-level performance, and can they be understood and applied to other countries?

These questions, along with a review of the historical, economic and systemic factors that contribute to the evolution and the coordination of the two policy frameworks within countries are considered at length in this report.

C. NOTE ON THE CHOICE OF REGION AND COUNTRIES

All regions of the developing world are currently coping with the issues of policy choices and policy coordination highlighted in this report. This report focuses exclusively on the African region where industrial development through technological change has become more pressing than ever before (UNCTAD, 2015). The recent economic performance of African countries is an embodiment of hope for the region. Not only must growth be sustained but more productive growth also needs to be promoted to retain these encouraging economic trends (See Rodrik, 2014).

The country selection criteria were based on three sets of parameters:

(i)   The developmental and institutional circumstance represented by the country:

While Nigeria is a developing country with a natural commodity (oil), Ethiopia is a least developed country (LDC) with a resource-concentration in agriculture. This is juxtaposed with the experience of the United Republic of Tanzania – which is an LDC with a mix of resource-based commodites and other sectoral activities. As a result, each of these countries serves to illustrate a developmental challenge in the realm of coordination of industrial and innovation policies for developmental outcomes.

(ii)   The ongoing policy transformation in industrial and innovation policies:

All the three countries discussed in this report have national vision documents, new industrial development strategies and STI policies that embody the aspiration of its leaders and policymakers to transform their nation into 'middle-income' economies within the next two to three decades.

(iii)   Difficulties faced in channeling R&D expenditure and GDP growth rates towards   technological learning:

All three countries have experienced relatively impressive GDP growth rates over the past decade, if not longer, and increased R&D expenditure as a percentage of GDP in the 2000s. Despite this, they have faced difficulties in focusing these investments into greater technological learning, particularly at the firm-level, as demonstrated by the lack of greater exports of medium- and high-technology products (see figures 2.1 and 2.2, and chapter II).

D. METHODOLOGY

A questionnaire designed by UNCTAD for this report was administered to firms and organizations in all three countries. The country surveys were designed to capture the interaction between innovation policies and industrial development strategies in recent decades; the surveys also seek to identify whether these policies impacted on addressing institutional and policy gaps or coordination structures, or systemic behavioural incentives (existing habits, practices, informal norms), and if this impact was felt by the firms (in terms of improved performance) at the level of the firm and the industry as a whole. The data collection was accompanied by extensive field interviews with key stakeholders, along with detailed reviews of the historical evolution of relevant policy frameworks on the industrial development and innovation policies in these three countries.

(i)   Technological intensity and sector coverage:

Different sectors need different kinds of technological skills, and often vary in intensity. By definition, low-technology intensive sectors do not call for much technological know-how, as opposed to sectors that are medium- or high-technology intensive which call for a wider, more versatile set of skills and know-how that are sector and industry specific. In order to capture these differences, three sectors were chosen to understand how firms learn and compete in sectors, which embody different levels of technological skills and know-how; these sectors comprise agro-processing (as an example of a low technology intensive sector); pharmaceuticals and health care, and the ICT sector, which embody both medium and high-technological intensity activities.

(ii)   Firm size and coverage:

The national surveys tried to cover firms of all sizes, as represented in the economy of that particular country. Firms were classified as small, medium and large based on standard classification. Firms employing less than 10 persons are regarded as micro-enterprises, while firms employing 10-49 and 50-199 persons are classified as small- and medium-scale enterprises, respectively. Those employing more than 199 are considered to be large-scale firms (Lall et al 1994, Oyelaran-Oyeyinka, 1997a).

(iii)   Focus of country-based work:

A variety of methodological tools was used to address the three basic lines of inquiries in this report, namely mechanisms underlying policy formulation and design, policy coordination during implementation, and policy impact on the firms.

For the first-level inquiry on the historical, economic and systemic factors that contribute to the way STI and industrial development policies are introduced in countries over time, extensive secondary and national level analysis was used to examine policy changes since the 1960s, in addition to policy documents from national archives. Survey interviews focused on eliciting historical experiences of a wide range of national stakeholders to understand how policy priorities were chosen, and the ramifications of these processes on economic performance of firms and sectors over time.

For the second-level inquiry on what factors determined and prevented the emergence of coordinated policy implementation and supporting institutional infrastructure, policy documents, survey questions and face-to-face interviews were used.

To understand how these factors impact firm-level performance in countries, i.e. the policy impact on firms and sectors, survey questionnaire and interviews with firms and other stakeholders were used as the main sources of information for the analysis.

The questionnaire was designed to capture the nature of innovation and the intensity of technological activities at the firm-level (i.e. to determine whether it consisted of process and product innovation, whether it was incremental, adaptive or R&D-based, and what forms of technological inputs went into production, marketing and feedback mechanisms). The survey questionnaire contained detailed questions on the activities of the companies, e.g. distribution, supplying, service provider and innovator, sources of innovation, nature of interaction, etc. It also sought to map the common venues of learning that local firms tapped into, existing modes of collaboration, general industry characteristics (capacity utilization, R&D capacity, export and import issues, etc.), the extent of public sector support, and policy incentives and institutional variables that impact firm-level performance. Questions in the survey also focused on issues related to the general performance of the relevant industrial sector and how that affected activity at the individual enterprise level.

(iv)   Additional data sources:

The analysis contained in each of the chapters also relies on secondary national and international databases. National income and broad sectoral output figures are taken from UNCTADstat database, while detailed sectoral statistics are obtained from national sources. ILO LABORSTA database is used for sectoral employment figures. UNCTADstat and Comtrade are the two main sources of aggregate figures mentioned in this report. National aggregate FDI figures were obtained from UNCTADstat, sectoral FDI flows in the world and developing countries were calculated using UNCTAD World Investment Report 2014 database. Structural statistics on physical and knowledge infrastructure are drawn from the World Bank's 'Ease of Doing Business', World Bank WDI, UNESCO database and the WIPO database.

E. DEFINITIONS

The terms industry, industrial policy, STI policies and innovation are multifaceted in nature, and are used in several ways in theory and practice. For the purposes of this report, they are defined in the following ways.

1. Industrial policy and industry

The term 'industrial policy' has undergone several different iterations over the past century, if not longer, and continues to evolve over time. Broadly denoting policies that promote economic restructuring (Rodrik, 2004), the term is used often synonymously with 'industrialization' policy or industrial development policy/strategy, which refers to the process of promoting industrial output in an economy (Lall, 1990, Warwick, 2013). In this report, the terms 'industrial policy' or 'industrial development policies' are interchangeable.

For purposes of the analysis, industrial policy is defined as the sum total of governmental actions undertaken to orientate and control the structural transformation process of an economy. Within this perspective, the analysis of production processes is the focus of investigation and review. This is consistent with the literature on the topic, where one of the accepted definitions is that industrial policy is "any type of intervention or government policy that attempts to improve the business environment or to alter the structure of economic activity" (Naudé, 2010). This broad definition of industrial policies covers all policy interventions that affect the performance of all sectors of the economy (see Cimoli, Dosi and Stiglitz, 2009).

The term 'industry', as used in this report, is broad and in keeping with the evolutionary nature of the concept, and is used to denote manufacturing, utilities, construction and mining, i.e. all industrial activities. This is in keeping with the way the sector

is defined and computed by most national statistical offices in the African region.

2. Innovation

Innovation is often confused to mean inventions, or the result of 'state of the art' R&D. This report employs a broader definition of innovation that is applicable to development. Innovation is considered the ability to develop new products/ processes/ organizational forms, which although may not be new to the world at large, is new to the local firm and the local context. This definition draws from Schumpeter's original works (1934, 1942), and is considered to be most relevant in the study of incremental learning and innovation capacity building in dynamic contexts (see Lundvall, 1993; Nelson, 1987).

3. Science, technology and innovation policy

The term 'innovation policy frameworks' (a term often used synonymously with STI policy frameworks or STI policies) refers to a purposive policy framework that is put in place to foster knowledge creation, adoption and distribution within a country, with an explicit focus on interactive learning among firms, public and private organizations that support innovation processes (Oyeyinka and Gehl Sampath, 2009).

F. REPORT'S CONTRIBUTION AND STRUCTURE

The report has embarked on what is normally a difficult exercise, namely, the collection of primary data based on a semi-structured questionnaire from firms in three African countries. This exercise is important because official data on firm-level activities is not always possible to obtain in several African countries, including those under consideration in this report. Other data that is often easily available in other countries, such as employment at the firm-level, R&D investment, total annual sales, etc. were not available given many of the sectoral firms under study were small and medium-sized enterprises or firms operating on the fringes of the informal sector.

As a result, the report brings to light new data and information, albeit of a descriptive nature, on how policies impact on firm-level behaviour in countries, and what factors matter in connecting policy to economic results. It lends evidence to the challenges and opportunities that countries face when coordinating innovation and industrial policies and incentives aimed at better supporting their enterprise sector. Many findings in the report are not entirely contextual – they are equally applicable to other countries in the African region, as well as more generally to other developing countries – and these are highlighted in chapter VI.

This report is structured as follows. Chapter II begins by elaborating the synergies and potential overlaps between innovation and industrial policy frameworks. It proposes a set of guiding principles that could help countries to align these policy frameworks to promote sustainable industrialization.

In chapters III to V of the report, the innovation and industrial policies, their day-to-day implementation and impact on the industrial sector of three African countries are examined. Chapter VI of the report combines the in-depth insights from the country-level investigations, with the principles highlighted in chapter II based on the analytical framework and the overall review of industrial policy and innovation policy frameworks in the African region. On this basis, chapter VI presents detailed findings on what matters in the industrial policy-innovation policy interface. The report concludes with relevant policy recommendations on what countries could do to promote innovation-led industrial development in an efficient and concerted manner.

NOTES

1. See Amsden (1989), Amsden and Chu (2003), Johnson (1986 and 1999) and other recent works on industrial policy and economic catch-up, such as Cimoli, Dosi and Stiglitz (2009) and Naudé (2010).

LINKAGES BETWEEN INNOVATION POLICIES AND INDUSTRIAL DEVELOPMENT

2

CHAPTER II
LINKAGES BETWEEN INNOVATION
POLICIES AND INDUSTRIAL DEVELOPMENT

A. INTRODUCTION

Interventions to support the growth of industry across sectors and promoting learning and innovation can be complex and take varied forms. The production of knowledge entails negative externalities due to its non-rivalrous and non-exclusive nature; this occurs as no individual or firm has the incentive to produce knowledge since it can be shared at marginal costs and is difficult to exclude. Market mechanisms, financial flows, transfer agreements, production processes all embody some level of negative externalities due to the complexities of trading information, or designing contracts with incomplete or asymmetric information. But at the same time, knowledge production and sharing also embody positive externalities that benefit society at large. Society benefits from the creation of new knowledge, and in addition, if collaboration within the private sector were to take place it would lead to a positive externality for learning and innovation for a large number of actors, which has added benefits for society over longer periods of time.

Ideally, governments need to take into account these positive externalities and provide an enabling environment for knowledge based-learning, in order to replicate the success of one or a few firms or sectors on an economy-wide scale (Stiglitz, 2014). Policies that address the replication of positive externalities of this kind are often more complex to design and implement than those that simply address negative externalities, thus explaining the difficulties that governments often experience when promoting innovation-led industrial development.

The fundamental purpose of industrial policy is to deal with market failures of all these kinds that impede industrial development; these failures not only concern labour allocation, credit institutions and the availability of goods, but also relate to knowledge accumulation and dissemination and new knowledge creation (see for example, Rodrik, 2007). In this endeavour, it also deals with several aspects of STI policy through incentives and instruments. Available policy instruments for industrial and STI policies are often applicable in both cases, and policy experience shows that they can be provided under either regimes. Examples include common industrial infrastructure for firms and sectors, industry parks, special economic zones (SEZs) and enterprise support.

Despite these overlaps and the complementary nature of both policy frameworks, neither of them is redundant, and close coordination is crucial to enforce developmental outcomes. For example, while industrial development strategies set overall economic targets, innovation policies provide the institutional infrastructure for learning, individual targets and supportive incentives to firms. While industrial development strategies aim to develop high-technology sectors, stimulate job growth and eradicate poverty, the sectors that will be prioritized and the modus operandi for such prioritization is usually set out in STI frameworks. While the emphasis of industrial development strategy of a given country may be on job growth or to facilitate recovery from a recent economic and financial crisis, the STI framework determines how this job growth can be based on technological development and on how high-quality and sustainable jobs can be created.

However, these linkages are not always automatically evident as they rely on a synchronization of policy goals and outcomes, as well as the calibration of well-established policy and institutional implementation mechanisms. This chapter therefore addresses the analytical framework and relative roles of each of these two policies and their common areas of overlap. It then proposes a set of principles to coordinate and promote the role of policies, which together represent an engine for growth, as well as help create a learning base for industry growth and expansion, as opposed to advocating for governmental interventions that simply seek to correct market failures (Stiglitz, 2014).

B. TRIGGERS OF INDUSTRIAL DEVELOPMENT

Industrial development depends on four critical factors: (i) skills development; (ii) technological change; (iii) industrial organization; and (iv) the ability to create a good business environment (UNIDO, 2013). Within these broader contours, national industrial policies are adjusted to accommodate several local, contextual or national concerns. The following goals of the Action Plan for the Accelerated Industrial Development of Africa summarize the key concerns of industrial policy as:

(i)   Accelerating industrial growth by promoting infrastructural changes;

(ii)   Developing human capital and resource mobilization for industrial development;

(iii) Fostering STI policies; and

(iv)   Developing a sound legal and institutional environment.

Similar industrial development goals are espoused in several regional and national policies and strategies across countries and regions globally.[2]

1. Why focus on industrial development policies?

The justification for industrial policies is controversial due to differing views on the role of governments and free markets within countries, and more recently, the world economy. A discussion on industrial policies invariably begins by questioning the need for such policies; indeed, one of the questions that are often asked is whether industrial policies or industrial development policies are needed in the first place.

Such policies are generally justified on the basis of two types of failures in information and coordination. Information failures arise when entrepreneurs need to access information on what in-demand products can be produced at relatively low costs in order to facilitate the move into newer production activities or sectors. The information needed to facilitate this includes accessing technological information and other firm-level inputs, which would lead to the profitable production of products and increasing market access. In practice, however, these many forms of information failures occur on a routine basis in developing countries; this impedes the ability of firms to access the information they need to diversify and profit from new economic activities, hence justifying governmental intervention (See also Lin and Chang, 2009). Coordination failures occur as a result of the difficulties that countries periodically face in coordinating the range of investments needed to promote large-scale industrial activities or projects. These investments are not simply of a financial nature but also concern human skills, technology, or plant-specific technological inputs – all key requirements if new sectors are to expand or grow. Given that developing countries face constraints generated by both types of market failures, industrial policies can serve as appropriate policy interventions to address these shortcomings.

Over the past half a century or so, industrial policy has been interpreted in various ways and shaped by different schools of thought. Aside from the two forms of failures, several other economic explanations have been advanced, particularly those based on Marshallian economics and the need to derive economies of scale, as well as explanations based on the need to increase productivity, especially labour productivity (see also UNCTAD, 2014). From the 1960s and up until the 1980s, industrial development strategies and policies were simultaneously understood as import substitution strategies that promoted an inward-looking, national economic development agenda (see for example, Dervis and Page, 1984). Over time, however, policy experiences of countries showed that governmental failures also often occur as governments are not clear about what is required to address industrial growth in their local contexts. Policies tend to become more effectively formulated over time, owing to policy experience and learning in countries, and despite varied levels of governmental failure, the role of the government remains fundamental because "…[i]ndustrial restructuring rarely takes place without significant governmental assistance" (Rodrik, 2004). Hence, the choice is not between one or another action, but rather revolves around how a government can complement market forces to achieve particular outcomes (Rodrik, 2004; Moudud, 2010, 2011).

2. Creating a supportive environment for industry

The government's role in industrial policy is to promote the accumulation of physical and human capital investments and to transform these investments into industrial learning activities by eliminating information and coordination failures (see Nelson and Pack, 1999; Lall and Teubal, 1998). Countries have adopted different approaches in implementing industrial policies or a broader industrial development strategy aimed at creating a supportive environment for industry. In Europe, for example, national industrial policies place more emphasis on some aspects than on other. Countries such as Sweden, Finland and Norway have relied on policies to build an extensive knowledge structure through technological progress, while concurrently employing other policies (see for example, Bairoch and Kozul Wright, 1996; Chang and Kozul-Wright, 1994; Chang, 2007), while others employed horizontal policies aimed at overall competitiveness (e.g. Germany), others still had a sectoral focus (e.g. France) (see Aiginger and Böheim, 2015).

Similarly, countries in the developing world have experienced their fair share of diver-

gences. For example, East Asian economies/countries concentrated on building a strong technology capacity and networks, as in the case of Japan (Fabiani, 2004), whereas Latin American countries used incentives to control or use direct foreign investment to promote technology transfer.

But on the whole, four different areas of intervention seem fundamental to the industrial policy/ industrial development strategy type of endeavour, namely: (i) improving technical and technological efficiency in firms; (ii) promoting enterprise/ business support; (iii) supporting industrial organization; and (iv) promoting a broader economic development strategy. These are briefly discussed below.

a. Improving technical and technological efficiency in firms

Entrepreneurship and diversifying production structures is a risky undertaking anywhere, but particularly so for firms in developing country contexts. The risks taken are different from those when engaged in R&D, and more specifically relate to the lack of information on what new products or processes could be produced given their technological capacity and existing market demand; how the inputs could be sourced; what forms of assistance is available to organize production efficiently; and how markets could be accessed. As a result, such risks play a critical role in decisions because if firms fail in their efforts, they alone bear the costs of this failure; even through their success is highly important from a social perspective.

A core focus of industrial policy has been to target support infrastructure for the development of new products or new processes. Laying the foundations for improving technical efficiency or adherence to quality standards involves indispensable factors, such as:

(i)   Ensuring the availability of scientific skills for R&D, as well as for production. These include tertiary education, vocational training and skills creation for industry support.

(ii)   Promoting production standards: These comprise requirements and specifications that firms need to achieve at all levels of the production process (i.e. plant and machinery specifications, output quality requirements, and marketing and delivery standards), as well as other different standards that firms need to comply with in order to be able to capture export markets.

b. Promoting enterprise/ business support

Industrial activity, particularly in resource-constrained contexts, calls for coordinated investments in large-scale projects, improved infrastructure, networking and other forms of business support (Macmillan and Rodrik, 2011). The provision of these wider systemic components are not necessarily in the interest of individual firms, but are essential in promoting the emergence of a competitive, collaborative business environment. These include the:

(i)   Provision of finance instruments;

(ii)   Promotion of enterprise development schemes, including SMEs, larger companies and state-owned enterprises (SOEs) that undertake the risks of production in new technological domains;

(iii) Creation of public utilities, such as uninterrupted access to electricity, physical or knowledge infrastructure;

(iv)   Policy instruments that promote an entrepreneurial culture, improve the business environment by fostering private sector partnerships and business incubation.

Direct government involvement in private sector training needs can take the form of support schemes, such as locally organized capacity building and training workshops and sponsorship support for entrepreneurs to attend international seminars and conferences. In addition, training opportunities can be provided at government research centres, public universities, technology centers and intellectual property offices.

c. Supporting industrial organization

Concentration of industrial activity often assists in the transition from low value-added activities to higher skill and technology intensive production (Rodrik, 1988; Mckormick, 1999). Facilitating specialization and creating scale effects of industrial production can be achieved through industrial clusters, industry parks, and the creation of industrial zones/ or creation of SEZs.

Industrial and exports processing zones have largely been policy-oriented initiatives by governments aimed at: (i) assembling the necessary infrastructure for enterprise development; (ii) provide incentives to both local and foreign investors; (iii) simplifying administrative procedures by establishing one stop-shop administrative offices; (iv) attracting FDI; (v) ensuring technological and knowledge spill-overs and learning among enterprises; (vi) facilitating specialization and scale effects; and (vii) driving nationwide industrialization processes and accelerating economic growth and development (Amirahmadi and Wu 1995; McIntyre, Narula, and Trevino 1996; Johansson, Helena, and Nilsson 1997, Madani 1999). SEZs or EPZs can comprise a range of specific export-led incentives for firms, including tax holidays, access to tax-free remittances, government grants and special loans facilities. SEZs are an important industrial policy tool but the results have been mixed across different developing countries. While countries such as China, the Republic of Korea, Malaysia and Singapore have been very successful, SEZs in sub-Saharan Africa are still facing challenges in creating employment and trade opportunities.

The key challenge in this regard is in creating linkages and spillovers, whereby the stronger linkages between the various actors based in a clustering area, industry park or SEZ are supported by a suitable business environment helps to promote knowledge and network spillovers.

d. Promoting a broader economic development strategy

Industrial development policies are often articulated and coordinated in such a way that their overall focus is on per capita GDP growth, as this variable is used to capture industrialization. This is based on the conventional understanding that industry value-added, as achieved by a gradual increase in manufacturing exports, have a positive correlation with real GDP per capita growth. However, these policies are also seen as crucial for broader economic development of countries seeking to link industrial development and associated income growth to a better distribution of employment opportunities and social outcomes.

It is therefore imperative that industrial policy is closely linked not just to higher per capita GDP ambitions, but also to policies that can improve the local business environment, as well as wage and labour policies and policies aimed at promoting social inclusion.

C. GOALS AND INCENTIVES IN INNOVATION POLICIES

Depsite the four broad areas of focus of industrial policy discussed above, increasing industrial productivity relies on the acquisition and mastery of a range of technological domains (Imbs and Wacziarg, 2003). As mentioned in chapter I, innovation policies are premised on the realization that simply supplying science or technology inputs (such as scientists or engineers) is not sufficient to enable such mastery; it is therefore necessary to support the emergence of a broader innovation ecosystem that supports linkages and collaborative networks amongst various actors. Accordingly, the underlying emphasis is not just on providing skills, knowledge infrastructure on greater R&D inputs, but also on facilitating the process of how these factors interact and result in greater technical efficiency in production processes at the firm-level.

1. Why focus on innovation policies?

Innovation policies, in the same manner as industrial policies, are also justified on grounds that coordination failures continue to persist within countries, which often do not allow the emergence of an enabling innovation environment. Economic historians and policy analysts studying the catch-up experiences of countries single out the lack of policy focus on innovation within industry, and the failure to promote technological capabilities, as major obstacles to competitiveness and growth in developing countries (see Chang, 2001; Lall et al., 1994).

Coordination externalities in providing an enabling innovation environment are to be expected given the wide range of factors that are usually required to promote innovation capacity and the time lag between when investments are made and the time when they can be expected to have an impact on promoting innovation. For example, investments in secondary and tertiary education facilities take a decade or two to result in superior human skills. The creation of R&D centres of excellence, or knowledge-based clusters for innovative firms, especially in high-technology sectors, also require investments to be made at least a decade before they bear fruit. These time lags makes it difficult for policymakers to foresee the technological outcomes of certain investments and plan/allocate resources, and to ensure continuity and consistency in efforts as required to achieve these longer-term outcomes. A focus on innovation policy is therefore emphasized as a guiding framework to enable such longer-term investments.

2. Policy objectives of innovation policies

The objectives of innovation policies are often defined broadly and not clearly articulated within one policy document, but within an umbrella framework of numerous policies on, among others, education, R&D, S&T and IPRs. Formulated as an array of policy initiatives aimed at promoting innovation and firm-specific learning, innovation policies seek to identify systemic failures. The latter failures are understood as being broader than just market failures and relate to why systems do not function holistically, and which missing incentives could be deployed to alleviate the difficulties in searching, acquiring, using and trading with information. Systemic failures also deal with a wide range of aspects, including interactions, collaborations and the role of non-economic actors in promoting innovation, and aligning industry technology needs with national development priorities. Broadly speaking, innovation policies seek to address these shortcomings by:

(i)   Fostering the technology absorption capacity of firms and other actors in the innovation system to increase their ability to benefit from knowledge flows; and

(ii)   Creating an overall innovation system by eliminating many of the systemic failures and promoting interactive learning.

a. Fostering technology absorption capacity

The first objective of innovation policy is to foster greater technology absorption capacity within the local ecosystem. Technology absorption capacity refers to the intrinsic ability of firms (and other organizations) to absorb existing knowledge and technologies to adapt or create new knowledge. These processes of incremental learning are fundamental to enhancing the efficiency and competitiveness of their business operations, and to promote technological change within firms, sectors and systems. Different kinds of learning help to build competences that play a role in how firms are able to adapt simple technologies (which may be relatively easier), or even progress to adapting sophisticated technologies, which may require other competences, including R&D.

The ability of firms in developing countries to transition through these stages and carry out more R&D-intensive activities depends on both endogenous and exogenous fac-

tors. Endogenous factors relate to the firm or the system itself and includes elements such as: the skills base; finance opportunities; collaboration venues; knowledge flows; in-house technological learning capabilities (in terms of number of skilled workers); training and retraining opportunities for workers; and mobility between university-industry. If channeled appropriately, exogenous factors regularly enhance the ability of firms to learn. Such factors include trade or technology licensing opportunities available from local and foreign sources; international quality standards that local firms may have to adhere to; opportunities to integrate global value chains interalia by producing value-added products; and benefits from technological spillovers arising from FDI. Both sets of factors impinge upon the ability of firms to engage in technological learning and production.

Building and sustaining technological absorption capacity therefore depends on strengthening all factors that jointly assist in the exercise. Existing data on how firms learn and channel technological information into production lends strength to the conclusion that firms engage in R&D when learning competences are well developed, and *in the presence of other factors*, such as skilled labour, a well-functioning system of public research, firm-level capacity to engage in production of medium- or high-technology intensive products, local and export demand for such production activities, and the availability of finance.

Therefore, it does not depend directly on how large national R&D investments are, but rather how the increased R&D spending is channeled into strengthening the systemic factors that promote technology absorption, such as public research capacity, university education (especially tertiary education), centres of excellence, and specific R&D incentives to increase university-industry collaboration, etc. Figures 2.1 and 2.2 plot the relationship between the share of GDP that is invested into R&D and the ability of countries to export, in order to illustrate this.

Figure 2.1 captures this relationship for medium technology-intensive exports (that

Figure 2.1:   Relationship between R&D expenditure (as a percentage of GDP) and exports of medium- technology intensity[3]

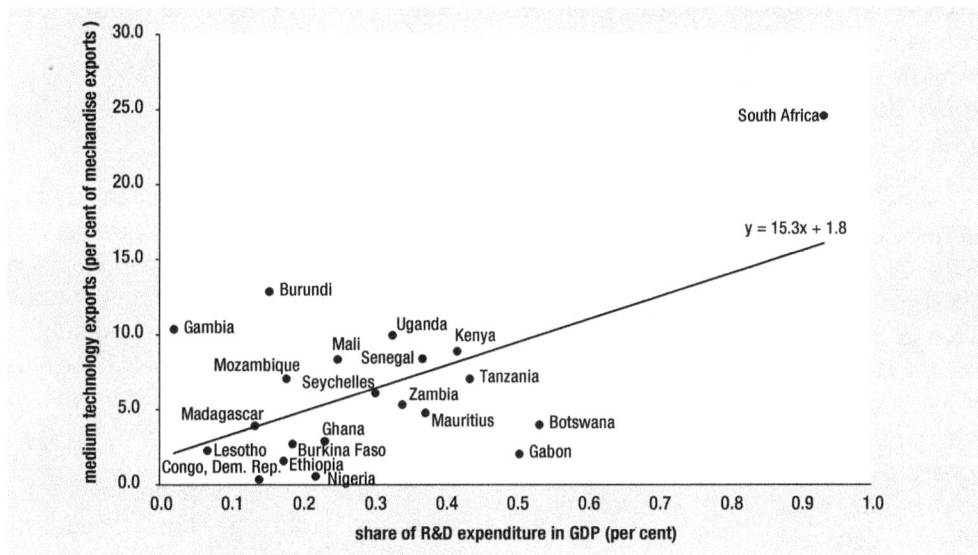

Source: UNCTAD calculations based on UNCTADstat (trade figures, accessed on 20 October 2015) and
         World Bank World Development Indicators (R&D figures, accessed on 20 October 2015).

is, products that call for medium-technology intensity), whereas figure 2.2 focuses on high-technology exports of countries. These two categories are presented here because product and processes that are either medium- or high-technology intensive call for some level of R&D expertise within firms. Figure 2.1 shows that although several countries are spending higher

amounts of GDP as R&D, this spending does not translate into greater amount of exports of medium-technology intensive products. The same is true in figure 2.2, helping to make the point that R&D investments results in greater technology-based exports only in the presence of absorption capabilities, which are shaped through a range of endogenous factors.

Figure 2.2:   Relationship between R&D expenditure (as a percentage of GDP) and technology exports (high intensity)[4]

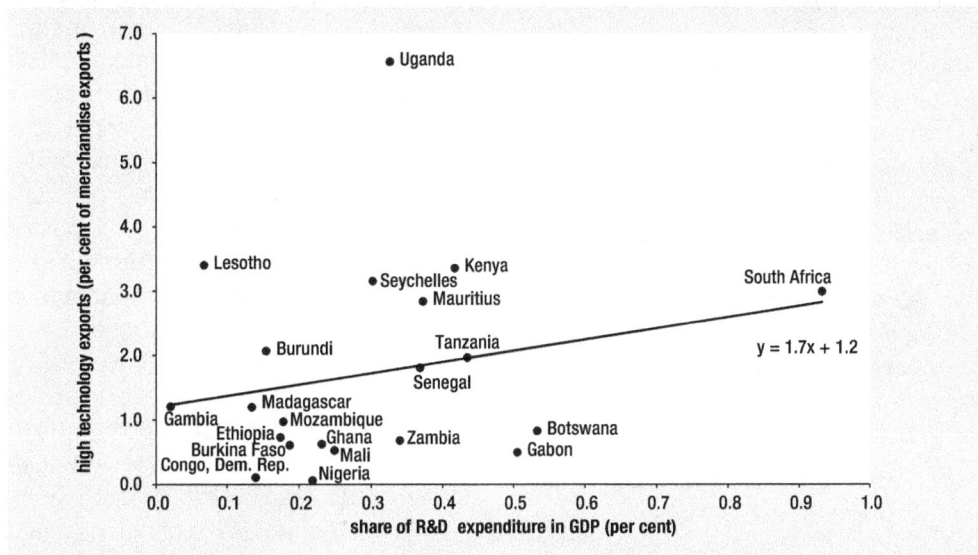

Source: UNCTAD calculations based on UNCTADstat (trade figures, accessed on 20 October 2015) and
         World Bank World Development Indicators (R&D figures, accessed on 20 October 2015).

Factors that shape technology absorption capacity are often contingent on developmental contexts. Researchers (see for example, World Bank, 2011) consider that low technological absorption capacity can be traced back to, *inter alia*:

(i)   Weaknesses in the process of skills creation:

On the supply-side, formal education curricula and training programmes are often not aligned with industry technology needs, creating a mismatch between what firms actually need and what graduates possess in terms of knowledge and qualifications. Such a misalignment can occur for two reasons: a gradual erosion of university standards, or lack of curriculum updates due to resource constraints, or a disconnect between university and industry. On the demand side, there could be a general lack of interest in skills development within firms, as a result of their focus on low-technology intensive production activities. These developments hinder the pace of technology absorption and competitiveness of firms.

(ii)   Lack of technological emphasis in trade:

Even though firms import machinery and other business equipment, workers are often unsure how they should be operated or maintained due a lack of interaction with technology experts, suppliers and chain intermediaries. Paucity of additional training beyond what is provided in the user manuals renders the mastery and application of machinery and equipment difficult and costly over time. Skilled personnel are often hired from advanced countries to help with the basic installation of imported machinery and equipment, and are called upon for help when equipment malfunctions. When this happens firms incur additional costs for the upkeep of equipment, instead of focusing on mastering new technology, thereby losing business competitiveness.

(iii)   Lack of industry-research collaborations and linkages:

The technology absorption capacity of firms can be greatly enhanced when they conduct joint R&D projects with domestic and international research organizations.

(iv)   Barriers to trade and FDI:

Lack of quality physical infrastructure makes the movement of goods and trade difficult, especially the movement of bulky machinery and equipment. Barriers to FDI inflows also influence technological absorption capcity, including: (i) the high cost of doing business; (ii) market uncertainties; (iii) the cost of acquiring business licences; and (iv) permits and scarcity of skilled labour.

To address these weaknesses, many developing countries have sought to strengthen education, as well as technical and vocational training institutions as part of their technical skills development process. In this context, it is critical to tap into all sources of relevant knowledge that may be held by specialists or institutions, particularly among those that can share tacit know-how, such as technology experts brought in by firms to help install new machinery, equipment and train local staff. For early stage-innovators or start-ups, trade and learning from capital goods could be crucial, as could gaining additional information on IPR or patents, as well as customer feedbacks (see figure 2.3 below). For growth-stage and maturity-stage innovators, R&D, technology licensing, research collaborations, technology transfer and sourcing of and learning from highly skilled technology experts could be important. Finally, since weaknesses in industry-research linkages could be attributed to weak or divergent research orientations, greater alignment between research institutions and applied and industrial research would help to make research more relevant, but would contribute to enhanced university-industry linkages. Other measures include facilitating the inflow of appropriate technologies and addressing infrastructural constraints to trade.

Figure 2.3:   Technological learning, technology flows and technological absorption: Factors and feedback loops

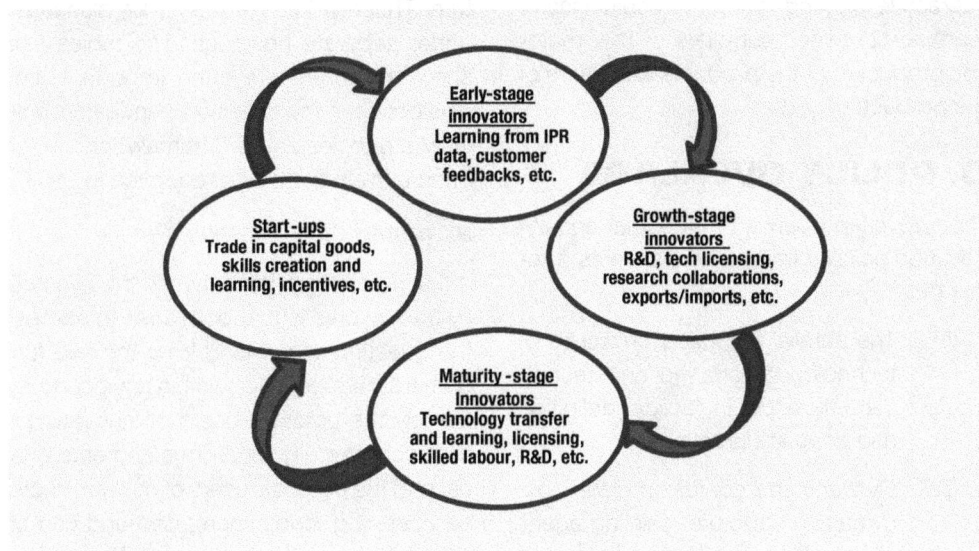

Source: UNCTAD.

b. Creating an overall innovation ecosystem

Creating an overall innovation ecosystem calls for the development of wide-ranging and dynamic relationships between innovation actors and related entities and supporting structures. Traditionally, countries have sought to promote innovation ecosystems which encompass and link all supply-side actors, e.g. knowledge institutions and demand-side actors, such as industry and supporting actors, such as the government, other financiers and global actors.

In general, the success of STI policies in providing an enabling environment depends upon a range of parameters, including:

(i)   High-level policy governance:

The absence of high-level policy governance has often led to misalignment and lack of clarity or poor interpretation of policy objectives at different levels of the governance structure. Also, there are instances where the roles of coordinating agencies overlap and duplication of efforts can lead to a waste of resources. Ensuring horizontal and vertical policy co-ordination and implementation is closely linked with the governance structure and function of a regulatory framework. Many countries have therefore moved to establish high-level policy governance to help deal with past policy coordination weaknesses.

(ii)   Simplifying administrative processes:

Common problems include bureaucratic hurdles faced by firms in obtaining documents, resistance to change, lack of computer know-how, and general lack of desire to adopt simpler processes, e.g. e-governance. To fix these administrative drawbacks, some countries have started migrating to cost-effective e-solutions that generate less paper work.

(iii)   Creating and strengthening linkages among innovation actors:

Many countries have sought to enhance linkages between industry and academia through collaborative research and spin-offs. Emphasis has also been placed on the importance of aligning academic programmes with industry needs as formal and informal networks are increasingly being viewed as part of the solution. Other relevant actors advocate for interactive learning, citing the usefulness of feedback between suppliers of raw material, producers of the products and users of the product.

D. POLICY OVERLAPS

Policy overlaps between industrial and innovation policies frameworks arise as a result of:

(i)   The parallel emphasis on issues of technological change and technical efficiency in policy definition and conceptualization;

(ii) Overlaps and confusion in the policymaking structure, existing agencies and mandates, poor business emphasis and business knowledge in policy processes; and

(iii) A lack of effective performance measures that could take stock and remedy missing coordination.

This section discusses the key overlapping functions of the two policy frameworks that are essential for industrial development, and is followed by an enumeration of the coordination issues resulting from such overlaps, and presents a set of principles for extracting synergistic results.

1. Overlapping domains of interventions in policy definition

Tables 2.1 and 2.2 contain a review of regional and national initiatives on industrial policy and STI policies in Africa. The review shows that there are a plethora of current initiatives. This is a welcome development and these initiatives often elaborate similar objectives. The synergistic impact of these various initiatives will only be felt if they are implemented in a coordinated manner in their respective national contexts. However, slight contradictions exist in the way goals, objectives or the policy implementation apparatus are defined. These relate to the following important areas: stimulating demand; innovation finance and investment

promotion; technological learning; and provision of supportive industry infrastructure. These will subsequently be discussed under separate headings. The tables also show that there are often important time lags between the definition of industrial policies/ strategies and STI frameworks, which often explain the lack of coordination.

a. Stimulating demand

A first area of overlap between the two policy frameworks is that both seek to address the question of demand from the two fundamental stakeholders within any economy, namely the potential users of innovation in the economy, e.g. local firms and enterprises, and the potential users of new products/ services, e.g. consumers. Demand can be stimulated when the government champions and makes a direct investment into certain sectors and technologies. To champion new sectors, industrial development schemes have typically often include direct state involvement, in the form of SOEs that seek to pioneer local production capacity in particular sectors. Over time, to ensure the efficiency of SOEs, countries such as Japan have often adopted state schemes aimed at improving functions at all levels of enterprises. In many other countries, privatization of existing SOEs (at a later stage of development) has also been a way to help promote enterprise development. Engaging in joint ventures and PPPs with some level of state participation could be a means to enhance enterprise development, and have been used by many of today's emerging economies.

Many African industrial development strategies provide incentives for state-sponsored investment which replicate the experiences of a number of developing countries in a variety of sectors (see Mazzucato, 2013), e.g. in the pharmaceutical sector. In Uganda, for example, a large public pharmaceutical company has been set up through a PPP between Quality Chemicals (Uganda) and Cipla Pharmaceuticals (India), with the help of state investment (see UNCTAD, 2010).

Table 2.1:	Industrial development initiatives in Africa

Pan-African initiatives	
	The Conference of African Ministers of Industry (CAMI) was initiated in 1971 to provide a platform for industrial policy dialogue.
	The Lagos Plan of Action (1980-2000) was fashioned to promote industrial development in Africa.
	The Abuja Treaty of 1991 was designed to harmonize the regional economic and social policies and promote regional production structures and infrastructural development.
	The Cairo Agenda of 1995 was launched to enhance industrial competitiveness of Africa through economic diversification.
	The 2001 New Partnership for Africa's Development (NEPAD) and the African Productive Capacity Initiative (APCI) were adopted in 2004 as part of the African-wide sustainable industrial development strategy.
	The African Union Conference of Minsters of Industry in 2007 drafted an Action Plan for the Accelerated Industrial Development of Africa.

Subregional Initiatives	
COMESA	The COMESA common industrial policy aimed at promoting manufacturing activities among Member States was presented and discussed during the 34th Meeting of the Inter-Governmental Committee (March 2015, Addis Ababa Ethiopia).
EAC	The East African Community Industrialization Policy, 2012-2032 is aimed at the structural transformation of the manufacturing sector through high-value addition and product diversification, based on comparative and competitive advantages of the region. EAC Member States are Burundi, Kenya, Rwanda, the United Republic of Tanzania and Uganda.
ECOWAS	The West African Common Industrial Policy was adopted in 2010 by the ECOWAS Authority of Heads of State and Government with an implementation vision up to 2030.
SADC	The SADC Industrial Policy Framework was adopted by the SADC Committee of Ministers of Trade (CMT) in June 2009 to promote the Industrial Upgrading and Modernisation among Member States (Angola, Botswana, Democratic Republic of Congo, Lesotho, Madagascar, Malawi, Mauritius, Mozambique, Namibia, Seychelles, South Africa, Swaziland, the United Republic of Tanzania, Zambia and Zimbabwe.

National Initiatives (selected countries)	
Angola	Angola has enacted a National Development Strategy aimed at, among others, the building of a national knowledge economy.
Egypt	Egypt's Industrial Development Strategy Industry: The Engine of Growth (2005-2025) was crafted in 2004 with the goal of transforming Egypt into a major manufacturing centre by 2025.
Eritrea	Eritrea with the support of UNIDO started working on an Integrated Industrial Policy for Sustainable Industrial Development and Competitiveness in 2004.
Ethiopia	Ethiopia adopted its Growth and Transformation Plan (GTP), 2010/11-2014/15 to: (a) build an economy with a modern and productive agricultural sector and enhance the capacity of the technology and industrial sector to assume a leading role in the economy; (b) sustain economic development and secure social justice; and (c) increase the per capita income of citizens to match the level of those in middle-income countries.
Gabon	Gabon's 2011 Industrial Policy is aimed at turning Gabon into an emerging economy by promoting 'Green Gabon', 'Industrial Gabon' and 'Service-Industry Gabon'.
Gambia	The Government of Gambia since 2010 has sought to formulate a National Industrial Policy (NIP) to establish conditions required by the private sector to maximize gainful employment at ever increasing levels of productivity within the framework of a sustainable environment, social justice and equity.
Ghana	Ghana launched its National Industrial Policy (NIP) in June 2011. The NIP is aimed at facilitating the country's industrialization agenda.
Lesotho	Lesotho adopted a National Strategic Development Plan (NSDP) 2012/13-2016/17 as part of the National Vision 2020 developed in 2001. The Plan is aimed at promoting a medium-term implementation strategy for, among others, MSMEs and the manufacturing sector.
Liberia	Liberia launched its 'Industry for Liberia's Future' in 2011 to accelerate the development of a thriving and competitive industrial sector, so as to become a middle-income country by 2030.
Madagascar	Madagascar issued its Industrial Policy Letter (2007-2012) to transform the country from a (predominantly) subsistence economy into a dynamic industrial economy that is strongly integrated into the global economy and to achieve socioeconomic development.
Mauritius	Mauritius launched its Strategic Plan for Industry, 2010-2013 to promote the manufacturing sectors, particularly small and medium-sized businesses.
Morocco	The Government of Morocco in 2014 launched a new Strategy (2014-2020), which is principally aimed at 'increasing the cadence of the Moroccan Industrialization'. The government also set up an Industrial Development Fund with a budget of $ 2.5 billion.

Namibia	Namibia launched its industrial policy in 2012. A supplementary document, the Industrial Policy Implementation and Strategic Framework details the targets, strategies and action plans on industrialization during the Fourth National Development Plan (NDP4) period, starting in the fiscal year 2012/13. Namibia aims to become a high-income industrialized country by 2030.
Nigeria	Nigeria launched its Industrial Revolution Plan in January 2014 with the goal of adding about NGN 5 trillion (or $ 25 billion) to annual manufacturing revenues in the next three to five years.
Rwanda	Rwanda launched its Industrial Master Plan, 2009-2020, in December 2009 in order to achieve global competitiveness.
Senegal	Senegal launched its Politique de Redéploiement Industriel (PRI), or Industrial Redeployment Policy, in 2005 with the aim of redistributing industrial facilities (currently concentrated in the Dakar region) across the country; re-orient the productive base towards promising sectors; and strengthen managerial capacities required to promote highly productive competitive industries (Cissé et al., 2014).
South Africa	South Africa's Industrial Policy Action Plan Economic, 2013/14-2015/16 is aimed at preventing industrial decline and supporting the growth and diversification of South Africa's manufacturing sector.
Swaziland	Swaziland has, since 2012, sought to formulate industrial policy for Swaziland to promote the development and growth of the manufacturing sector.[5]
Tanzania, United Republic of	The United Republic of Tanzania adopted its Integrated Industrial Development Strategy 2025 in December 2011. The goal is to promote agriculture-led and resource-based industrialization.
Uganda	Uganda drafted its National Industrial Sector Strategic Plan, 2010/11-2014/15 in 2009 to follow through with the implementation of the objectives of the 2008 National Industrial Policy Framework for Uganda's Transformation and Competitiveness. The policy vision is to build the industrial sector into a modern, competitive and dynamic sector fully integrated into the domestic, regional and global economies.
Zimbabwe	Zimbabwe's Industrial Development Policy (2012–2016) seeks to maximize revenue deliverables from the exploitation of natural resources through the enhancement of investment in industrial sector.

Source: UNCTAD.

Table 2.2:	STI policy initiatives and strategies

Pan-African initiatives	
	The first African Ministerial Conference on Science and Technology (AMCOST) under the auspices of AU and NEPAD was held in November 2003 in Johannesburg, South Africa. AMCOST is a Specialized Technical Committee of the African Union that promotes pan-African STI policies and programmes. Ordinary meetings of the AMCOST are held once every two years, with the provision for extraordinary meetings when necessary.
	The Second African Ministerial Conference on Science and Technology (AMCOST2) was held in Dakar, Senegal, from 27 to 30 September 2005. At this meeting, the delegates adopted the Africa's Science and Technology Consolidated Plan of Action (CPA). The CPA articulates the Africa's commitment to developing and applying STIs to enable Africa to harness and apply STI for poverty eradication and sustainable development.
	The Extraordinary Conference of the African Ministerial Council on Science and Technology (AMCOST) was held in Cairo, Egypt, from 20-24 November 2006. Delegates deliberated on STI issues, including a proposal to establish an African Presidents' Committee for Science and Technology, as well as a proposal for the African Strategy for Technology Transfer and Acquisition of Domestic Technological Capabilities.
	The Third Ordinary Session of the African Ministerial Conference on Science and Technology (AMCOST III) was held in Mombasa, Kenya on 12-16 November 2007. Deliberations focused on a draft, consolidated framework on the protection of traditional knowledge, intellectual property, individual and community rights.
	The Fifth Ordinary Session of the African Ministerial Conference on Science & Technology (AMCOST V) took place on 12-15 November 2012 in Brazzaville, Congo. The session focused on the strategies and the implementation of Africa's Science and Technology Consolidated Plan of Action (CPA).
	The Ministerial Forum on Science, Technology and Innovation took place in Rabat, Morocco from 14-17 October 2014. The forum was organized by the African Development Bank, following the first STI ministerial conference, which was hosted by the Government of Kenya in 2012. The forum was designed to raise political awareness of S&T in Africa and promote youth employment, human capital development and inclusive growth.

Subregional initiatives	
COMESA	The Bureau of the Council of the COMESA Ministers responsible for STI inaugurated the COMESA Innovation Council on 8 April 2013, in Kampala, Uganda to enhance S&T in the region.[6] The COMESA Innovation Council is tasked with providing advice to member states on existing new knowledge and innovations and best means of introducing them in the region.
EAC	The EAC produced its Development Strategy, 2011/12-2015/16, to strengthen efforts to develop regional industrial R&D, technology and innovation systems. The strategy specifically seeks to invest in higher education and training, technology development and innovation in the EAC region.

ECOWAS	The Second Conference of ECOWAS Ministers for Science and Technology was held on 24 March 2012 in Yamoussoukro, Côte d'Ivoire adopted the ECOWAS policy on S&T (ECOPOST) and related action plan. The Commission also plans to create a Directorate for STI to play a key role in the socio-economic development of the region.[7]
SADC	SADC ministers responsible for STI approved the SADC Science, Technology and Innovation Strategic Plan 2015-2020 in June 2014 in Maputo, Mozambique. The plan is aimed at promoting the development of STI in the region through regional coordination, institutional development, policy harmonization and resource mobilization. The policy also seeks to promote transfer and mastery of STI within the region.[8]

National initiatives (selected countries)

Angola	The Angola National Policy for Science, Technology and Innovation was adopted following the passing of a Presidential Decree No. 201/11 in 2011. This STI policy was developed to complement the country's development strategy, and is aimed at building a knowledge society, combat poverty and improve quality of life.
Botswana	The National Policy on Research, Science, Technology and Innovation was launched in 2011.[9]
Burundi	The National Policy on Scientific Research and Technological Innovation was drafted in 2011,[10] but the policy was revised and launched, along with its implementation framework, in August 2014.[11]
Cameroon	The National Policy for the Development of Information and Communication Technologies was launched in September 2007.[12]
Egypt	Egypt launched the Decade for Science and Technology 2007/16 Strategy and introduced the Developing Scientific Research Plan 2007/16 (OECD, 2014). There is also the national ICT Strategy covering the period between 2012 and 2017.[13]
Ethiopia	Ethiopia launched its National STI Policy in 2012 to alleviate poverty and transform Ethiopia into a middle-income country by 2023. The National STI plans seek to develop capabilities in the country so as to enable rapid learning, adaptation and utilization of effective foreign technologies by 2022/2023.
Ghana	The National Science, Technology and Innovation Policy was drafted in 2009 and adopted in 2010. The STI policy along with its Vision 2020 aims to: (a) transform the country into a middle-income country; (b) create endogenous S&T capacities that are appropriate to national needs, priorities and resources; and (c) create an S&T culture to address the country's sociocultural and economic problems. The policy also specifically aims to combat global warming by increasing the use of renewable energies and allocating minimum of 1 per cent of GDP to support the S&T sector.
Kenya	The National Science, Technology and Innovation Policy and Strategy were launched in March 2008.[14] Science, Technology and Innovation Act (2013), Draft National Science, Technology and Innovation Policy (2012).
Lesotho	Lesotho's Science & Technology Policy (2006-2011) was implemented in 2006.[15]
Malawi	The Government of Malawi passed the Science and Technology Act (2003) and later established the National Commission for Science and Technology (NCST) to promote, support, coordinate and regulate the development and application of science, technology and innovation in order to create wealth and improve the way of life of the people.[16]
Mauritius	Mauritius has a Draft National Policy and Strategy on Science, Technology and Innovation (2014-2025).
Mozambique	Mozambique approved its STI Strategy in 2006. The strategy aims to: (a) promote STI in industry and public sector, promote technology transfer; (b) stimulate the use of ICT for good governance and service delivery and for the diffusion of knowledge; and (c) support human resource development in STI.
Namibia	A National Research Science and Technology Policy was formulated in 1999 and the subsequent enabling act was adopted in 2004.[17] The Ministry of ICT drafted the Information Technology Policy for Namibia in 2008.[18] A Draft Innovation Framework Policy (2011) has been tabled for final implementation. (see UNU-MERIT, 2015)
Nigeria	The National STI policy of 2011 was reviewed and the revised policy was launched in 2012.
Rwanda	The National STI policy (2006) was revised in October 2014 but has yet to be approved by the Cabinet.
South Africa	The ICT R&D Strategy for South Africa was finalized in 2007 and is being implemented under the auspices of the Information Society and Development Plan (ISAD) Plan of South Africa. The National Research and Development Strategy was published in August 2002, and the Department of Science and Technology published the Ten-Year Innovation Plan (2008-2018) in 2007.[19]
Tanzania, United Republic of	The STI policy was reviewed between 2008 and 2013. A new revised policy is expected to be implemented in early 2016.
Uganda	The Cabinet approved its first national STI (NSTI) policy in 2009. The government launch in 2012 of the National STI Plan 2012/2013 - 2017/2018 was aimed at supporting the implementation of the 2009 NSTI policy.
Zambia	Zambia adopted the National Policy on S&T in 1996 (Daka and Toivanen, 2014). The 1996 Science and Technology Policy was reviewed in 2008, and the process of drafting a new national S&T policy also began in 2008.[20]
Zimbabwe	Zimbabwe's STI Policy was launched in 2012.

Source: UNCTAD, based on UNU-MERIT (2015) and national sources.

Another way is for the government to simply list specific sectors as priorities and identify a range of incentives for their promotion; this sends a signal to both consumers and companies that the sectors or specific sets of technologies are supported. Such instruments do not directly involve governmental entrepreneurship (through SOEs) and are elaborated to stimulate demand in both industrial and innovation policy frameworks, but particularly often figure in innovation frameworks. These include R&D grants and loans, industry grants, prizes, tax credits and government procurement. Government procurement is acknowledged to be an essential tool for stimulating new technological knowledge, particularly in certain sectors of the economy (see Georghiou, 2014). Procurement also serves as a tool to stimulate innovation in two important ways, namely: to foster innovation in key growth sectors (such as pharmaceuticals and health care); and to foster innovation in particular sectors of the economy that are critical for sustainable development (e.g. renewable energy). SEZs (see section b) are yet another mechanism to facilitate firm-level activities based on export demand.

b. Finance and investment

A second area of overlap is in the area of finance and investment promotion. Finance is often the single most crippling bottleneck to firm-level activities, and access to innovation finance is a major problem for firms of all sizes in developing countries, particularly in Africa. Industrial policy incentives, such as duty-free importation of capital goods and raw materials for selected products, exports tax exemptions, as well as exemptions from levies imposed on exports have been the hallmarks of import-substitution policies in the past. Different schemes exist that enable entrepreneurs with limited resources to start businesses, including: Government loans with flexible repayment plans; loan guarantee programmes for firms; and trade credits that allow enterprises making bulk purchases to pay in one, two or three months' time. Examples of agencies that are sometimes

established under industrial policies are the National Competitiveness Councils, which deal with enterprise financing. Other policy instruments to address finance bottlenecks include mid-term loans for industry, as well as microcredit schemes (either directly through banks or as facilitated through governmental programmes).

However, since most government-sponsored enterprise support schemes either have a direct or indirect impact on firm-level R&D and production activities, they overlap with innovation policies, which also often focus on financing schemes, both through direct and indirect instrumentalities. Incentives such as R&D grants, credits or subsidies often perform the dual role of stimulating demand and attenuating finance bottlenecks.

c. Accelerate technological learning through an enabling environment

Another fundamental concern that is common to both policy frameworks is technological learning. A review of country-level policy frameworks shows a concomitant focus on promoting skills creation and business and technical advisory services.

From a normative perspective, however, there are (or ought to be) differences in the focus of skills creation in the two policies *per se*. While industrial policy theoretically aims to create skills infrastructure for the development of new products or processes, particularly those aimed at improving technical efficiency or adherence to quality standards, the supply of engineers and scientists have long been the focus of S&T policies (see King, 1991). In some ways, the focus of industrial policy is to balance the supply of skilled scientific personnel, while at the same time taking into consideration the supply side of the equation, i.e. the kinds of knowledge and skills that are required at the firm/enterprise level to increase the technical efficiency of production. At the enterprise level, it is important to have access to skilled personnel capable of dealing with day-to-day enterprise operational needs, and promoting routine learning-by-doing activities, such as design, prototyping and reverse engineering.

Table 2.3: R&D technicians and scientific researchers in enterprise development

Requirements	R&D technicians	Scientific researchers
What they are?	– Possess applied scientific skills – Support scientific activities	– Possess basic research skills – Lecture/ teach in science departments and academia
How are they qualified?	– Apprenticeship – Vocational training – College education	– Masters/ PhD degrees – advanced research skills
How they work?	– Develop products – Test for quality, health safety, standard protocols, etc – design, research and prototype at the plant level	– Conduct basic/ applied research in labs – Can be involved in product, process development
Where they work?	– Work in enterprises – Manufacturing activities	– Academic institutions – Research centres – Technology Laboratories – R&D laboratories of firms
Length of training	-Can be acquired in less than two years -Bachelor degree	– Masters or PhD degree – Several years of experience and continuous learning

Source: UNCTAD.

Such technical expertise is often distinct from the scientific research capacity to be found in laboratories. While both kinds of capabilities are essential for the creation of new products or processes, there are key differences in the domains in which they operate.

While R&D technicians possess applied scientific skills that are useful for industry, scientific researchers mostly carry out research (basic or industrial) in laboratory settings. R&D technicians support scientific researchers with the practical training skills they acquired in the course of an apprenticeship, vocational education, on-the-job training or interaction with industry. R&D technicians also operate at the product development and testing stages in the manufacturing sector. Scientific researchers in academic or R&D labs apply advanced research skills acquired during university studies, centres of excellence and specialized research environments. These differences are elaborated in table 2.3, and promoting learning at the industry level calls for both kinds of skills development.

In other words, in addition to the supply of scientists and engineers, a range of other technical and business advisory services are needed to channel such skills to plant-level R&D, production and industrial competitiveness. In order to enable this, business advisory services are usually aimed at the acquisition of improved and innovative skills to improve enterprise-level performance (see Turner, 2011), or technical advisory services focus on providing advice and coaching on specific innovation related skills and activities. This technical advice can relate to: specific in-house R&D activities; access to and exploiting technical knowledge; ICTs; technical information; meeting national and international quality standards; and other quality-related matters.

In practice, however, the distinction between the two policy frameworks is blurred, particularly with respect to which policy incentive is provided by which policy. The supply of scientists and engineers, business advisory services and technical advisory services are often neglected; or provided for in both frameworks. In addition, services, such as advice on specific in-house R&D activities, information on technical issues, and assistance in implementing specific international standards, are also often either neglected, or provided for in both frameworks, with low levels of coordination between the agencies offering the incentives.

Other overlapping incentives relate to the provision of knowledge infrastructure, such as increased R&D, centres of excellence, technology centres, which can either involve setting up new public research institutions and/or restructuring existing facilities to help cater for industry needs.

d. Establishment of supporting institutions

This refers to the so-called "intermediate organizations", such as business incubator units, technology incubators, S&T parks, SEZs or industrial parks, which all seek to promote the development of physical and knowledge infrastructure and the creation and strengthening of institutional linkages (box 2.1).

Both policy frameworks contain specific incentives covering industry parks, SEZs, EPZs, in addition to incentives for instruments that promote collaboration, e.g. grants to ensure that public research institutes work alongside firms, or university patenting to create incentives for industry-oriented research, and industry park or science park programmes. Other incentives that could facilitate feedback between suppliers of raw material, producers of the products and users of the product for interactive learning and further innovations are also mentioned in both policy frameworks, but often without the relevant feedback loops.

E. WHAT MATTERS

At a practical level, the manufacturing imperative that serves as the basis for industrial development strategies and plans will depend on how well these two policies are coordinated. Evidence and debate seems to be converging on a few essential factors. One of these factors is that industrial policy, or any intervention aimed at economic and industrial restructuring cannot function as as standalone framework if it is to succeed in promoting technological learning, or lead to sectoral change, diversification and social inclusion. Industrial policy needs to be closely calibrated with other policies, especially innovation policy. In fact, in recent times, scholarship has emphasized the need for innovation-oriented industrial policies, or a systemic innovation and industrial policy to guide development as a singular entity (Mazzucato, 2013; Aiginger, 2012).

A second, related result is that if industrial policy is to result in these outcomes, it

---

Box 2.1:   Industrial hubs, zones and parks

Industrial policies or strategies usually seek to promote instruments of agglomeration of firms, such as clustering. To faciliate this, incentives are provided to house suppliers, manufacturers and service providers, along with universities or centres of excellence performing research in the same or relevant areas. Clustering initiatives are centered on the idea of linkages and spillovers, whereby the stronger linkages between the various actors based in a clustering area, as supported by a suitable business environment, helps to promote knowledge and network spillovers, and promote economies of scale and scope.

SEZs or EPZs provide a one-stop shop for a range of specific export-led incentives for firms, including permits and licences, access to tax-free remittances, government grants and special loans facilities. In some countries, firms located in and exporting out of SEZs/EPZs are entitled to certain tax remittances as well. Despite their relevance as an industrial policy tool, the results are rather mixed across different developing countries. Countries, such as China, Republic of Korea, Malaysia, Mauritius and Singapore, have successfully established SEZs, whereas others, particularly in sub-Saharan Africa, are still facing challenges in creating employment and trade from their SEZs.

Similarly, industrial parks can assist firms to move from low to higher value-added activities, mainly due to the absence of specific industry infrastructure, such as those for technology incubation, or specific large-scale investments that cannot be made by individual firms. Examples include sectors where heavy public investments are needed for infrastructure industries (e.g. steel, cement), but also other sectors requiring particular kinds of knowledge or other physical investments. In the case of pharmaceutical sector, firms usually tend to benefit from a common active pharmaceutical ingredients (API) park, or common bioequivalence facility to assist local firms to test and comply with production standards. In many other cases, industrial parks often perform the basic, yet critical, function of providing industrial premises where a dependable basic infrastructure is available, e.g. uninterrupted power supply, warehousing facilities, and water for industrial activities.

Source: UNCTAD.

should focus on *getting the policy processes right* (Rodrik, 2004). Alignment issues arise because STI policies (or earlier S&T policies) were often conceived much later, or independently of other industrial development strategies. This leads to to two separate sets of institutions, overlapping agencies with similar mandates and policy processes that often do not communicate with each other or collaborate. These alignment issues need to be resolved to ensure that industrial and innovation policies are well-calibrated and synergistic in two ways. Firstly, policies need to be implemented in a coordinated way. Secondly, policies need to be evaluated on the basis of their impact on firm-level and

industry performance. Despite concerted efforts enterprise growth and expansion has remained a far more elusive goal for many countries because connecting the impact of policy on practice is difficult to measure. This is especially the case because a large number of systemic factors shape the impact of industrial and innovation policies on the firms that policy processes may not capture and fix, especially if the success of policies is not directly measured through firm-level performance at the ground level.

Thirdly and most importantly, an innovation and industry friendly climate is not about specifying the kinds of financing incentives,

Table 2.4: Industrial and innovation policies for development: Key alignment issues	
Industrial policy gaps	**Innovation policy gaps**
Gaps in policy articulation and planning	
Gaps in policy definition (no focus on linking to distribution, investment, etc) Industrial policy delinked from innovation policy/ technological learning Policies not clearly linked to roadmaps or implementation strategies Difficulties due to slow policy transitions Narrowly focused on the development of certain sectors to the neglect of others	Policies often not clearly articulated to focus on innovation Not aligned to industries' needs/ industrial strategies and national plans Excessive supply-side focused on S&T Lack of focus on technological absorption
Lack of policy coherence and policy competence	
Lack of micro-, meso-, and macro-policy linkages Incidence of overlapping mandates and jurisdictions among agencies Little or no articulation of regional or local priorities and contexts Poor planning Prevalence of standalone approaches with competing agency mandates Lack of proper implementation mechanisms Low or no provision for revisions based on feedback and/ or policy failure Human resource gaps leading to less than satisfactory policy outcomes	Weak coordination structures Incomplete/ competing agency mandates Unsustainable/ad hoc measures Low financing to implement programmes Lack of clear roadmaps for coordination Neglect of issues of institutional resistance and inertia Lack of policy competence to foresee overlaps Lack of investments in scientific and knowledge infrastructure
Resource use, resource constrains and duplication	
Inter-agency rivalry and competition for scarce resources due to overlapping mandates Unrealistic programmes with small or no budgets Lack of focus on market-driven opportunities Low consideration of funding in policy articulation Neglect of financial realities driven by ambitious projects Lack of focus on project/ programme success	Inter-agency rivalry and competition for scarce resources due to overlapping mandates Low consideration of funding in policy articulation Excessive focus on funding basic R&D as against funding industry R&D Lack of effective innovation financing strategies Duplication of incentives (with those already contained in industrial policy)
Insufficient capacity to conduct monitoring and evaluation	
Neglect and failure to learn from past institutional failures and successes Lack of skilled personal to conduct proper policy evaluations Lack of proper data and policy performance indicators Lack of good monitoring mechanisms Lack of regular follow-up	Neglect and failure to learn from past institutional failures and successes Lack of skilled personal to conduct proper policy evaluations Absence of good data Lack of proper data and policy performance indicators Lack of good monitoring mechanisms Lack of regular follow-up
Lack of coordination between policymaking, governmental interventions and business environment	
Policies are often not geared to local sectoral realities Industry policy frameworks not accompanied by industry and business support organizations Neglect of the needs of the private sector	STI policy frameworks not accompanied by industry and business support organizations Low focus on real firm-level hurdles and needs. Low focus on collaborative linkages and interactive learning

Source: UNCTAD.

but rather about identifying the activities, the beneficiaries in need of support (i.e. the kind of firms and what they should be focusing on), and how this support can be accessed by deserving firms (Foray, 2015; Aghion et al, 2012; Aiginger and Böheim, 2015).

The table below presents key alignment issues between industrial and innovation policies in a schematic form.

Some principles offer a good basis to synergize the two policy frameworks, and these are discussed here.

1. Identify and eliminate policy redundancies

As shown in table 2.4, some industrial and innovation policy gaps also relate to weaknesses in the policymaking structure (UNCTAD, 2003). The latter encompasses the institutional and departmental set-up, as well as rules and processes for policymaking and coordination (see Bianchi and Labory, 2008). Common weaknesses include: overlapping jurisdictions; industrial policy delinked from innovation policy; institutional weaknesses; unsustainable or ad hoc policy measures; budget constraints; lack of political will; and policy continuity. Getting the policy processes right calls for opening channels of communication between relevant agencies, actors, coordination and review of existing political support to implement policies, and change parameters as needed (Robinson, 2010).

These review mechanisms to eliminate policy redundancies in definition, design and process is largely explained by the relative success of industrial policies over the past five decades in countries/economies, such as the Republic of Korea and Taiwan, China, as compared to countries, such as Argentina in Latin America and countries, such as Ghana and Zambia in Africa (Robinson 2010).

2. Promote policy coherence and competence

At the national level, industrial and innovation policies are more effective when: (i) there is policy complementarity (Fukasaku

et al., 2005); (ii) when they are accompanied by clear and adequately funded budgets; and (iii) skilled employees that can implement the policies. Common industrial and innovation policy gaps at the national level are often due to the absence of micro-macro policy linkages, little or no articulation of an overall development vision in both policies (or inversely, the lack of implementable developmental vision), or a lack of proper coordination between policy incentives leading to a proliferation of standalone approaches.

Weaknesses in industrial and innovation policies are explained by the absence of policy competence (Meyer-Stamer, 2009). They can also fail when local priorities are neglected due to: (i) low or no policy learning; (ii) low or no policy revisions based on feedback and policy failure; (iii) low ability to assess lessons learnt; and (iv) human resource and technology gaps. Lack of policy competence also leads to poor implementation of even well-meaning policies in an industry friendly way, such as flexibilities of the national IPR regimes. Failure to follow through on investments in scientific and knowledge infrastructure development, unrealistic and unsustainable ad hoc programmes, or added transaction costs for local firms engaging in learning and innovation can also result in a lack of policy competence. As elaborated in the previous section, the lack of policy coherence is a critical issue, which often occurs in policy definition, implementation and task allocation and agency mandates in all the operative domains of both policies. Elimination of these redundancies in actual practice will be critical to ensure the effective implementation of the two policy frameworks.

3. Use resources carefully

In developing countries in particular, older policies often tend to be replaced by newer ones in an effort to address development. However, the mandates of existing agencies or newly created agencies are often not redefined in an appropriate manner, and nor is there a clear delineation of budgeting and performance monitoring measures (see next point). This not only creates inter-agency rivalries and competition for scarce resources, which goes against the fundamental objectives of both industrial and innovation policies (given that both intend to increase collaborative linkages and provide an enabling environment), but also leads to a duplication of scarce resources, resulting in the ineffectiveness of programmes, projects and incentive mechanisms.

4. Develop capacity for proper policy evaluation and monitoring

Industrial and innovation policies fail when there is lack of capacity to conduct proper policy evaluation and monitoring. Poor choice of monitoring indicators, timelines and evaluation techniques can compromise results and affect policy implementation. Deficiencies in information and data gathering processes can also affect industrial policy evaluation and monitoring. Lack of resources also impacts proper policy monitoring and evaluation.

5. Coordinate policymaking efforts and implementation with the local business environment more closely in order to engage the private sector

Finally, it is the institutionalized patterns of policymaking, governmental intervention and business-government relations that shape the process of industrial adjustment. In other words, what matters is not the identification of sectors of importance, or tools/mechanism for industrial organization (such as clustering, SEZs, or industry and science parks), but rather to recognize the relationship between policymaking (i.e. historically institutionalized patterns) that shape the way in which the policies are implemented

(the informal rules of the game) and the way business, and especially local businesses, react to it. Most of all, policymaking still tends, in large parts, to take place without paying much attention to the needs of the private sector. Changing the industrial and innovation performance of countries calls for a review of what went wrong (in terms of policy implementation, review and monitoring), and how these lapses can be avoided. Going ahead, engaging the private sector through policy action will be critical for overall industrial performance.

F. CHAPTER SUMMARY

This chapter has derived an analytical framework to assess industrial development and innovation policy frameworks, with the aim of showing the important areas of overlap between the two policies. Such overlapping, often contradictory, policy incentives lead to confusion in agency mandates, institutional redundancy and waste of scarce resources in developing countries. The chapter also shows that the overlaps and lack of implementation coordination is, in large part, caused by several historical, institutional constraints faced by countries. As shown in tables 2.1, 2.2 and 2.4, STI policies have often evolved independently and much later than industrial development policies, leading to difficulties in getting the policy processes right. In order to address these issues, the chapter derives a set of five principles that can assist countries to align the two policies, and minimize the negative impact of non-complementary linkages on industry and sectoral growth in their economies.

NOTES

1. The same aims are applicable across countries. The European Union's industrial policy, for example, envisages:
   (a) Speeding up the adjustment of industry to the need of structural changes;
   (b) Encouraging an environment favourable to initiative and the development of undertakings throughout the Union, particularly small and medium-sized undertakings;
   (c) Encouraging an environment favourable to cooperation between undertakings; and
   (d) Fostering better exploitation of the industrial potential of policies of innovation, research and technological development. See Article 173 of the Treaty on the Functioning of the EU (TFEU), the Europe 2020 strategy, and 'An industrial policy for the globalization era'. Available at: http://www.europarl.europa.eu/atyourservice/en/displayFtu.html?ftuId=FTU_5.9.1.html

2. As today's R&D spending can only stimulate future exports, the figure plots the R&D share of a country on its exports five years later. Latest available R&D figures for the years 2004-2008 and export figures for the years 2009-2013 have been used, when available, to prepare this figure.

3. As today's R&D spending can only stimulate future exports, the figure plots the R&D share of a country on its exports five years later. Latest available R&D figures for the years 2004-2008 and export figures for the years 2009-2013 have been used, when available, to prepare this figure.

4. http://www.trademarksa.org/news/swaziland-commerce-ministry-wants-develop-industrial-policy

5. http://www.comesa.int/index.php?option=com_content&view=article&id=659:innovation-council-inaugurated&catid=5:latest-news&Itemid=41

6. http://news.ecowas.int/presseshow.php?nb=086&lang=en&annee=2012

7. http://www.acgt.co.za/wp-content/uploads/2014/07/SADC-regional-cooperation-and-the-expectations-required-from-member-states_Anneline-Morgan.pdf

8. http://www.unesco.org/new/en/natural-sciences/science-technology/sti-policy/country-studies/botswana/

9. https://www.ist-africa.org/home/default.asp?page=news-doc-by-id-print&docid=9090&

10. https://www.ist-africa.org/home/default.asp?page=news-doc-by-id-print&docid=9090&

11. https://www.ist-africa.org/home/default.asp?page=news-doc-by-id-print&docid=9090&

12. Egypt Ministry of Communications and Information Technology National ICT Strategy: 2012-2017 http://mcit.gov.eg/Upcont/Documents/ICT%20Strategy%202012-2017.pdf

13. IST- Africa http://www.ist-africa.org/home/default.asp?page=ictpolicies

14. UNESCO, http://www.unesco.org/new/en/natural-sciences/science-technology/sti-policy/country-studies/ and IST-Africa http://www.ist-africa.org/home/default.asp?page=ictpolicies

15. Malawi National Commission for Science and Technology. http://www.ncst.mw/welcome-to-ncst/

16. Namibia National Commission on Research, Science and Technology. http://www.ncrst.na/about-us/Programme-Policies/47/

17. IST- Africa http://www.ist-africa.org/home/default.asp?page=ictpolicies

18. IST- Africa http://www.ist-africa.org/home/default.asp?page=ictpolicies

19. New Partnership for Africa's Development (NEPAD) http://www.nepad.org/system/files/032_DAY5_Presentation_on_STI_12_07_2012.pdf

COORDINATING INNOVATION AND INDUSTRIAL POLICY: NIGERIA'S EXPERIENCE

3

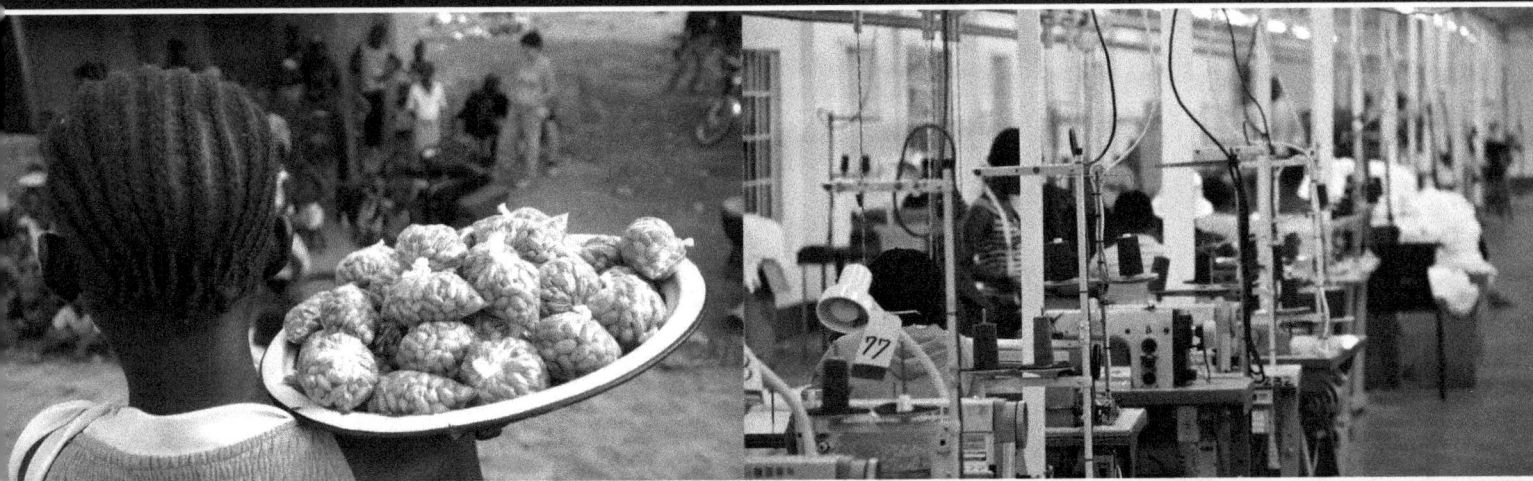

CHAPTER III
COORDINATING INNOVATION AND INDUSTRIAL POLICY: NIGERIA'S EXPERIENCE

A. INTRODUCTION

Nigeria is currently home to about a quarter of the population of sub-Saharan Africa and faces developmental challenges in ensuring that GDP growth translates into prosperity for all. Rising growth rates reflect the dynamism of the local economy, which is being increasingly acknowledged for its relatively open investment climate and improving infrastructure, both physical and institutional. In 2014, Nigeria conducted an exercise to rebase its GDP that resulted in the country being proclaimed the largest economy in Africa, surpassing South Africa by a wide margin.[1] Despite these new optimistic figures, Nigeria aspires to have a mature economy with a diversified industrial base, and to reduce reliance oil-based exports, which currently account for over 90 per cent of its export earnings.[2]

This chapter seeks to analyse the industrial development of Nigeria from a historical perspective, and will critically assess the role of innovation and technological learning in promoting Nigerian industry. The analysis is timely and highly relevant, particularly in the context of recent debates in Nigeria on the need for a more diversified economy, as well as the need to build on recent successes in industry. The analysis is also important from yet another perspective: an ever-widening divide has grown between the rich and the poor, as reflected in the growth of poverty levels from 24 per cent in 1980 to 66 per cent in 2010.[3] This raises very important questions of how industrial development can be made more equitable in Nigeria.

In accordance with the methodological structure outlined in chapter I, this chapter is based on a country-based field survey conducted for this report, which is clarified in box 3.1 below.

Section B presents an analysis of the underlying drivers of growth in the economy, tracing the challenges in structural diversification from the 1960s until now. The section then moves on to present the main policy and institutional framework in Nigeria for technology and innovation-led industry development, and assesses milestones in the development of both sets of policies over time. An analysis based on field survey results presents the day-to-day constraints currently faced by Nigerian firms, and how these are explained by limitations of the policy environment is presented in section C. Section D links these results to the overall challenge of promoting industry led growth in Nigeria, particularly from the perspective of resource-rich developing countries.

Box 3.1:	Scope and details of data collection in Nigeria

In 2013, an UNCTAD questionnaire was administered to elicit information from 245 firms in order to collect primary data on factors that affect industrial development and innovation capacity at the firm and sectoral level. Of the total of 245 firms, 200 questionnaires were retrieved. The firms were selected to represent specific industrial sectors, namely: ICTs, pharmaceuticals and health care and agro/food processing. A large number of interviews and surveys took place around Lagos and Ibadan, as well in other areas where there was a concentration of SMEs operating in these sectors. Firm size was chosen based on the overall profile of the sectors, in order to maintain representativeness.

Source: UNCTAD.

B. OVERALL TRENDS IN THE ECONOMY

The Nigerian economy has followed a rather complex trajectory since independence, accounted for in part by its over-reliance on oil-based exports and the premature openness of its economy. After independence in 1960, Nigeria's economy went through a period of appreciable growth, recording real GDP growth of 3.1 per cent annually in the first decade. The economy grew by 6.2 per cent between 1970 and 1978, according to data released by the Nigerian Bureau of Statistics. Growth rates have continued to be high since then, particularly when compared with other African economies. Between 1990 and 2014, the annual real per capita growth rate was 3.3 per cent, outstripping growth rates of most other African developing countries over the same period.[4] Nigeria's growth rate rose in the 2000s (2000-2014) and peaked at 4.6 per cent, which is about double the average growth rate of most other African developing countries for the same period (figure 3.1). [5]

1. Underlying drivers of growth

Industry, the second largest sector in Nigeria, accounted for about 26 per cent of GDP in 2013 (table 3.1). However, most of this was attributable to the oil sector: with mining and utilities accounting for 13.7 per cent of the national income. The reliance of the economy on crude oil exports, which accounted for about 70 per cent of total exports during the past four decades, led to a shift away from industrial activities of a productive nature, leading to low structural change, low dynamism and over-dependence on a single commodity.

This reliance on oil-based primary exports, which began in the 1970s also resulted in a dramatic shrinking of its agricultural sector over time. Statistics show that the agricultural sector, which accounted for about 27.1 per cent of GDP in 1970, shrank by almost one fifth (21 per cent) by 2000. Since the beginning of the new millennium, Nigeria's democratic government has attempted to secure overall macroeconomic stability and has managed to stabilize and reinstate some of the gains made in the agricultural sector. In 2004, the Nigerian government also enacted a National Economic Empowerment and Development Strategy (NEEDS), which is a medium-term development framework aimed at re-engineering growth and productivity. Some of the policy measures under this initiative have helped to stabilize some of the gains made in the agricultural sector (see Briggs, 2007).

Figure 3.1:   Real per capita GDP growth rate in Nigeria vis-a-vis other regions of the developing world, 1970-2014 (in per cent)

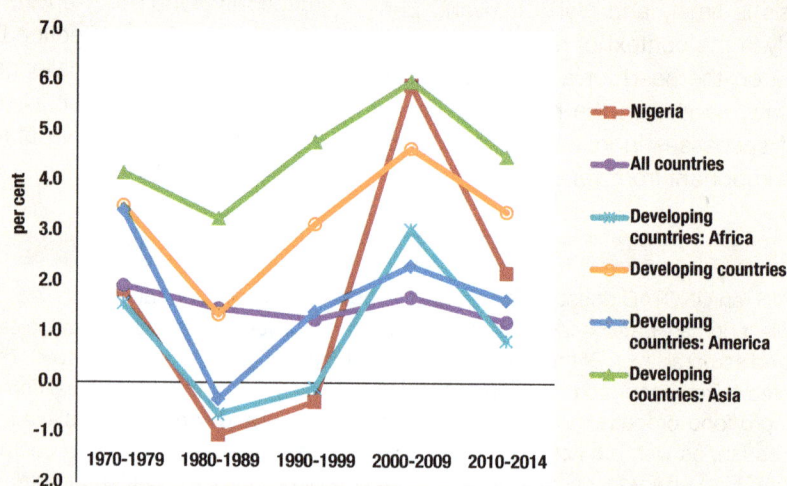

Source: UNCTAD calculations based on UNCTADstat (accessed on 20 October 2015).
Note: 2014 figures are estimates.

Table 3.1:   Distribution of Nigeria's GDP by sector, 1970 to 2013 (in per cent)

Sectors/ Years	1970	1975	1980	1985	1990	1995	2000	2005	2010	2013
Agriculture, hunting, forestry, fishing	27.1	18.8	17.4	23.2	24.6	27.0	21.3	25.6	23.9	21.0
Industry	19.3	24.8	28.5	18.2	25.8	25.1	29.9	23.7	25.3	26.0
Mining & utilities	5.0	7.9	12.0	7.9	17.2	18.0	18.7	14.9	15.9	13.7
Manufacturing	4.4	6.5	6.5	6.6	5.0	5.4	7.8	6.2	6.6	9.0
Construction	9.9	10.4	10.0	3.8	3.6	1.7	3.3	2.6	2.9	3.3
Services	53.6	56.4	54.1	58.6	49.6	47.9	48.8	50.7	50.8	53.0

Source: UNCTADstat (accessed on 24 September 2015).

In comparison with these two sectors, the services sector in the country's largest sector, accounting for 53 per cent of the total GDP as of 2013. This is quite similar to other developing countries in Africa where the services sector on average captures 46 per cent of GDP.[6] However, it should be noted that the relative increase in share of services is due to the rebased figures, which show a 26 per cent increase in services' share (according to 2012 data) to 53 per cent (following new data released after the rebasing exercise). What accounts for this difference is still unclear and may need to be re-assessed.

The employment effects of the three main sectors of the economy are rather telling. On the whole, despite its variable contributions to overall GDP over the past four decades, agriculture remains the largest source of income for Nigerian households and accounted for about 49 per cent of total employment in 2007.[7] The services sector is the second largest in terms of employment opportunities, accounting for about 43 per cent of total employment in 2007.[8] Industry, on the other hand, accounted for only 7.5 per cent of total employment in the country. Out of this, manufacturing accounted for 5.6 per cent, mining stood at 0.5 per cent, utilities at 0.7 per cent and construction at 1.6 per cent.[9]

2. Challenges for structural diversification: 1960s to the present day

While Nigeria went through a period of industrialization in the 1960s and the 1970s,

it experienced negative growth rates in the 1980s as a result of political instability and civil unrest. Several studies note that the introduction of the Structural Adjustment Programme (SAP) in 1986 initially reversed the lagging growth, and led to an annual GDP growth rate of 4 per cent between 1988 and 1997 (see for example, Agboli and Ukaegbu 2006; Brautigam, 1997). However, over the longer term the country underwent some degree of de-industrialization. This section will describe the underlying causes for these changes and place Nigeria's policy development on innovation and industrial development in a historical perspective by tracing the various policies that were introduced from the 1960s until the present day.

a. Nigeria's national development plans

Nigeria's earliest strategy for industrialization can be traced back to the Nigerian National Development Plan of 1962-68, which laid out a plan of import-substitution for industrialization. Although the plan itself sought to invest in some large-scale industrial infrastructure (such as a development bank and large-scale energy infrastructure), studies note that a focus on local industrial capacity and acquisition of technologies only emerged in the second National Development Plan of 1970-74 (Chete et al, 2014). This second plan recognized the need to promote technology acquisition to boost industrial activity. The government invested in large-scale iron and steel plants, petrochemical companies (in Eleme), fertilizer plants (in the Onu and Kaduna regions), oil refineries (in Port Harcourt, Warri and Kaduna regions) (see Oyeyinka, 2014).

However, Nigeria discovered oil during the second National Development Plan, which resulted in a lower emphasis on industrial development over time, as a result of the expansion of the oil sector. The third National Development Plan of 1975-80 envisaged greater public sector investment in industry, particularly heavy industries, and was implemented at a time when oil exports from the country were at an all-time high. Also around this time, the investments made in large-scale public sector industries in the first and the second national plans became unprofitable, partly due to the slow follow-up on these projects after the discovery of oil (Oyeyinka, 2014).

The failure of these public sector enterprises was accompanied by the acknowledgement that Nigeria's national developmental plans had not focused sufficiently on technological acquisition and creation of skills, and particularly on the managerial and implementation capabilities required at the plant level in large-scale industries. In an effort to rectify this lack of focus and support ailing public sector enterprises, some efforts were made to acquire managerial expertise from other countries and to send Nigerian nationals to public sector enterprises abroad to gather tacit know-how. However, these initiatives also failed since there was no capacity for technology absorption within the country.[10]

b. The 1998 National Industrial Policy

The 1998 National Industrial Policy was launched to resuscitate the country's industrial sector and formed part of the new national political agenda. The industrial policy of 1998 was aimed at structurally diversifying the national economy by promoting new sectoral activities, increasing the manufacturing value-added of products, as well as the use of greater local inputs to diversify the industrial base and promote exports. It also aimed at increasing the ability of private sector firms to participate in industrial activity. The policy was augmented by the National Industrial Master Plan, which provided the trajectory for the development of major industrial sectors.

The industrial policy framework contained several incentives aimed at encouraging industrial exports, some of which have yet to be implemented. The incentives included, among others, export credit guarantees,[11] export expansion schemes,[12] export price adjustment,[13] and subsidies for the use of local raw materials in export production.[14] In an effort to address this, the Federal Ministry of Commerce and Tourism offered several other incentives, such as the manufacturing-in-bond scheme, which aimed to assist potential exporters of manufactured products to import raw materials free of duty for the production of exportable products, and a supplementary allowance for companies that pioneered new products for exports.

Despite these efforts, little change occurred in productivity growth. For example, data from the Nigerian National Bureau of Statistics shows that the manufacturing sector's contribution to GDP was 5 per cent as of 1999, and that its share continued to fluctuate between 3.9 and 4.1 per cent between 2006 and 2010. By the end of the 1990s, a single commodity accounted for the majority of exports and the share of manufacturing and construction declined, with mining and utilities capturing an even larger share of GDP over this period, accounting for half of the GDP in 2000. These developments signalled the onset of the Dutch disease, and growth in the natural resource sectors began to hinder the development of the manufacturing sector, and in Nigeria's case, the growth of the services sector as well.

Around this time the Nigerian government established a number of institutions to stimulate the industrial sector, particularly manufacturing, as a subsector of industry, to reverse or at least mitigate the impact of the growth of the natural resource sectors on the development of the manufacturing and services sectors. Prominent institutions that emerged alongside existing line agencies included the Nigerian Bank of Industry, Nigeria Export-Import Bank, Nigeria

Export Promotion Council, Nigerian Investment Promotion Council and the Small and Medium Enterprises Development Agency. Some of these agencies emerged to support ongoing changes in the industrial sector. For example, the Small and Medium Enterprises Development Agency was created in the early 2000s to nurture SMEs, the largest group in the industrial sector. The government also put in place an updated Industrial Development Plan in 2014 (see next section).

c.   National vision statements: Nigeria Vision 2010 and 2020

Along with the industrial policy of 1998, Nigeria also adopted the Vision 2010, the purpose of which was to provide an overall national vision for development. The Vision 2010 report aimed to transform Nigeria into an African economic powerhouse, with a significant presence in the global economy by 2010. Vision 2010 set out specific annual GDP growth rate targets for the economy: The country was expected to achieve a growth rate of 9 per cent between 2001 and 2005 and 10 per cent between 2006 and 2010.

To reinforce the goals of the Vision 2010 report, the government more recently adopted the *Nigeria Vision 20:2020* (hereafter NV 20:2020), which is a long-term strategy aimed at transforming the Nigerian economy into one of the top 20 global economies by 2020. The Vision 20:2020 aims at a growth expansion of the Nigerian economy from $173bn in 2009 to $900bn by 2020 (with a per capita income of $4,000). The key policy milestones to achieve this transition NV 20:2020 included:

(i)   Maintaining an average annual GDP growth rate of 13.8 per cent;

(ii)   Reducing national inflation to a single digit figure (see NPC, 2010);

(iii)   Increasing the contribution of the manufacturing sector from 4 to 12 per cent of GDP during 2010-2013.[15]

The NV 20:2020 is now being implemented through a series of medium-term plans (called National Implementation Plans or NIPs), the first of which was developed for the period of 2010-2013. The second and third NIPs of 2014-2017 and 2018-2020, respectively, have also been formulated. In order to complement the first NIP, the Nigerian government introduced the Transformation Agenda (TA) 2011-2015.

Most recently, the Federal Ministry of Industry, Trade, and Investment with inputs from other government agencies and the private sector introduced an Industrial Revolution Plan in January 2014. The Plan also recognizes the importance of coordinating industry development with trade and investment regimes.

d.   Nigeria's National STI Policy

The failure of the Third National Development Plan led to an analysis of the causes behind stagnating industrial production. One of the main failings of the first three development plans is widely considered to be due to the lack of a comprehensive approach integrating technology acquisition and training in the day-to-day workings of public sector enterprises.

To address this, Nigeria adopted a S&T Policy in 1986 to address the difficulties faced by public sector firms, focusing mainly on technology acquisition and technology transfer issues. The policy was, however, not very effective in promoting technological know-how, particularly in Nigeria's ailing public sector enterprises. A systematic analysis of these failed projects, as reflected in studies (see Imevbore, 2001; Oyeyinka, 1997b) showed that in almost all cases, there was a lack of conceptualization of the process of technology capability and a missing focus on technological learning amongst enterprises beyond the generic acquisition of hardware machinery and equipment.

The 1986 S&T Policy was revisited in 1997 and again in 2003, through policy reviews that tried to place greater emphasis on coordination of the country's S&T system; set sectoral priorities, tackle the question of funding of S&T activities and emphasize upon collaboration in order to make it more effective. The 2003 policy review was based on the premise that Nigeria needed a policy framework that allowed for more systemic interaction to promote science and technology at the same time as innovation capacity.

In 2005, Nigeria embarked upon a system-wide review of its S&T framework along with UNESCO following the widely held view that the 2003 policy review was not comprehensive enough, particularly in creating a focus on innovation capacity. At the end of the 2005 review, Nigeria adopted a new national STI policy framework in 2011. The earlier 2003 document, as noted in the preamble of the national STI policy of 2011, is now considered to form a compendium of the main subsectoral policies at the core of the STI policy framework.

The mission of the national STI policy of 2011, as contained in Article 2.3, is to assist in "[e]volving a nation that harnesses, develops and utilizes STI to build a large, strong, diversified, sustainable and competitive economy that guarantees a high standard of living and quality of life to its citizens." The STI policy vision is in line with that contained in the Nigeria Vision 2020, namely to have a large, strong, diversified, sustainable and competitive economy by 2020. The specific objectives of the policy are rather elaborate, and include the creation of innovative enterprises, promote interactive learning, provide better funding, and provide an overarching institutional framework for STI.

C. INNOVATION AND INDUSTRY GROWTH: RESULTS OF THE FIELD SURVEY

1. Enterprise characteristics in the three surveyed sectors

Apart from varying technological intensity (see chapter I), there were some other reasons for choosing the three sectors for the survey. Firstly, both ICTs and pharmaceuticals are priority sectors in Nigeria, whereas the agro-processing sector is currently recognized to be one of the most successful economic sectors. Secondly, in order to be able to use the survey to gauge the impact of Nigeria's policy changes over the past two decades, companies established in the 1990s and early 2000s were chosen (to the extent possible) to understand the impact of institutional support (and changes therein) on industrial performance, R&D and innovation.

Thirdly, the survey sought to cover companies of all sizes in order to capture differences between firm size and performance. According to government estimates, as of 2010, there were about 17.3 million MS-MEs, contributing about 46.5 per cent of the country's GDP (NBS, 2010). In the firms surveyed, as of 2012, 22 per cent of the companies employed 1-9 staff (micro enterprises), 50 per cent of the companies employed 10-49 employees (small scale), 15 per cent of the companies employed 50-199 people (medium-scale), and 12 per cent were large-scale companies employing over 199 personnel on a full-time basis.[16]

2. Survey results: Nature of innovation in the three sectors

Given that firms engage in different kinds of learning activities, the questions targeted the number of product or process activities that the firms are engaged in, and whether these products and processes are new to the local firm (indicating routine learning), the local market (indicating incremental innovations, local adaptations), the regional or global market (indicating potential innovation inputs into new industrial products and processes). Other questions were directed at understanding the nature of firm-level learning.

a. New process and product innovations

Survey results show that the propensity of firms to engage in new product or process development depended on the technological intensity of the three sectors surveyed (table 3.2). The highest share of firms en-

Table 3.2: Distribution of firms carrying out new product and process developments					
	Health care/ pharmaceuticals	Agro-processing	ICT	Other	Overall
Number of firms[1]	52	79	44	25	200
Product development[2]	23/45	58/69	18/29	17/18	116/161
Share[3] (%)	51.1	84.1	62.1	94.4	72.0
Process development[4]	22/44	53/64	24/27	17/18	116/153
Share[5] (%)	50.0	82.8	88.8	94.4	75.8
Products new to[6]					
Firm	13/45 (28.8%)	21/46 (45.7%)	12/23 (52.2%)	9/10 (90%)	55/124 (44.4%)
Local market	2/45 (4.4%)	14/46 (30.4%)	8/23 (34.8%)	–	24/114 (21.1%)
Regional market	–	–	–	–	–
Global market	–	–	–	–	–

Source: UNCTAD calculations based on Nigeria Field Survey, 2013.

Note: The category 'other' in this and all other tables in this chapter indicates firms that participated in the survey, but could not be strictly classified under one or another category. For example, firms that engaged in agro-processing and nutraceuticals, or companies that provided ICT services, such as access to the internet.

[1] Number of firms that participated in the survey.
[2] Firms that carried out product development out of the total number of firms responding to this question.
[3] Share of firms that carried out product development out of the total number of firms responding to this question.
[4] Number of firms that carried out process development out of the total number of firms responding to this question
[5] Share of firms that carried out process development out of the total number of firms responding to this question.
[6] Number of firms that indicated product or process is new out of the total number of firms responding to this question.

gaging in product development was in the sector that called for the lowest technological capacity (agro-processing), 84.1 per cent of the firms were reported to engage in such activities; the percentage decreased in the case of health care and pharmaceuticals as well as ICTs (to 51 and 62 per cent, respectively). In the case of process innovations (which refer to creation/adoption of new processes), both agro-processing and ICT firms reported higher innovation rates when compared to health care and pharmaceuticals (50 per cent as opposed to over 80 per cent in the former two sectors).

To understand the nature of these innovations, the firms were asked to rate the novelty of their innovative efforts. As table 3.2 shows, a large number of firms surveyed admitted that their new process/product innovations were new to the firms, and that only 21 per cent of these firms admitted that their products were new to the local market. These survey results show that firms tended to produce products/processes that were new to the firm or local market in low technology sectors, thus reflecting the low level of technology intensity within firms.

The survey also studied the nature of the activities undertaken within the firms that led to new product/process development. While 30 per cent of the firms admitted to conducting some form of R&D, 65 per cent of the companies only focused on production and marketing, while 22 per cent reported that they were engaged in product development, and that 12 per cent were engaged in testing and laboratory services.[17] Firms in ICTs and pharmaceuticals reported a very small amount of contract manufacturing (2.6 per cent of their total activity). The survey results show that a majority of the firms were engaged in marketing and distributing products, rather than creating new products/processes.

b. Collaborations and sources of technological information for firms

There are several ways to determine innovation capacity at the enterprise level. First and foremost among these is the propensity to engage in R&D. Others include determining the sources of technological information/learning in the firms, or the nature of plant machinery and equipment that was used in the competitive industrial activity.

The survey looked at all of these aspects to understand the sources of technological learning.

A majority of the firms surveyed reported that they relied on their own efforts to remain competitive (table 3.3). While the highest rated factor was in-house R&D, interviews showed that the understanding of what constituted R&D at the firm-level was very different (see explanation in the next section). In addition to their own efforts, firms cited other important factors, including the support provided by intermediary organizations, collaborations with industry associations and 'others'. The category 'others' was elaborated as copying and reverse engineering, a practice undertaken by a large number of the surveyed firms. Figures in the table below are ranked in order of importance between 0 (not impor-

tant) and 2 (extremely important).The survey questionnaire sought to understand the contribution of various sources of technology to new product/process development at the firm-level. Figures in table 3.4 below are ranked in order of importance between 0 (not important) and 5 (extremely important). Skilled manpower was rated as the most important factor in new product/process development, followed by the quality of local infrastructure services, availability of venture capital, participation in local SMEs schemes, transfer of personnel between local firms/R&D institutions (for training) and so on.

On the whole, the survey showed that the majority of firm-level activity is still focused on sourcing spare parts and assembling them to produce products of relatively low technological intensity but as confirmed

Table 3.3:	New processes and organizational systems of firms by sources				
Technology sources	Health care/ pharma	Agro-processing	ICT	Other	Overall
In-house R&D	1.37	1.34	1.68	1.10	1.37
Support from intermediary organization	0.47	1.29	1.55	1.00	1.08
Collaboration[1]	0.53	1.29	1.48	1.00	1.07
Others	0.37	1.11	1.53	1.00	1.00
Licensed[2]	0.47	1.18	1.13	1.00	0.95
Adapted from competitors	0.45	1.09	1.23	0.75	0.88

Source: UNCTAD calculations based on Nigeria Field Survey, 2013.
Note: Figures only include firms that reported new process development. Figures represent the mean of rankings between 0 (not important) and 2 (extremely important).
[1] Within industry association.
[2] From technology supplier.

Table 3.4:	Contribution of various factors to new product or process development				
	Health care/ pharma	Agro-processing	ICT	Other	Overall
Scientific/skilled manpower	3.43	2.72	3.42	3.68	3.31
Quality of local infrastructure services	2.98	2.71	2.36	3.42	2.87
Availability of venture capital	3.28	2.62	2.32	2.95	2.79
Participation in local SMI development schemes	2.67	2.06	2.29	3.16	2.54
Intellectual property protection	2.98	2.28	1.88	2.74	2.47
Transfer of personnel to local firms or R&D institutions (for training)	2.56	1.97	2.03	2.84	2.35
Participation in joint government/firm technology transfer coordination councils	2.34	2.10	2.14	2.74	2.33
R&D collaboration with local institutions focused on R&D	2.72	1.69	2.03	2.05	2.12
Collaboration with local universities on R&D	2.50	1.56	2.20	2.21	2.12
Government incentives for innovation	2.23	1.86	1.81	2.37	2.07

Source: UNCTAD calculations based on Nigeria Field Survey, 2013.
Note: Figures represent the mean of rankings between 0 (not important) and 5 (extremely important).

by face-to-face interviews, respondents viewed and reported the sourcing and assembling of these spare parts as R&D. A large number of the surveyed firms were concerned about how to source spare parts and equipment for the day-to-day functioning of the enterprises, which also led to low utilization rates (see next section).

3. Survey results: Sectoral weaknesses, innovation constraints and industry performance

The survey also focused on understanding innovation constraints and industry performance issues that affected the activities of the firms. The results are presented here in four separate categories: (a) failings in the general innovation environment; (b) issues of competitiveness; (c) policy impediments to learning; and (d) the lack of collaborative linkages.

a. Failings in the general innovation environment

Surveyed firms identified several difficulties related to learning. One of the largest problems they faced was the lack of support to engage in technological learning within the national innovation system.

i. Knowledge related issues

A major prerequisite for successful innovation is the competence of available human capital. Amongst the firms interviewed, only 6.9 per cent of staff had obtained a PhD, while 25.1 per cent of staff had either a Bachelor's or Master's degrees. Most other employees did not have a higher academic qualification. As a result, a large number of the firms reported to be limited by a lack of skilled personnel.

As captured in the next section, firms also identified the issue of weak linkages between SMEs and large firms and knowledge centres.

ii. Physical infrastructure related issues

The number of telephone lines rose significantly from a mere 750,000 lines in 2001 to over 171.9 million in September 2013; over

the same period, teledensity increased from less than 5 per cent in 2001 to 86.6 per cent. However, the ubiquitous lack of efficient physical infrastructure, such as electricity for industrial purposes continues to be a drag on to the economy (see next section).

b. Competitiveness-related issues

The survey found that Nigerian companies continue to face many of the same innovation infrastructure deficiencies that local enterprises faced in the 1970s and the 1980s. For example, prior to 2001, most Nigerians lacked access to telephones as a result of outdated technologies and the inability of NITEL Plc (the then public monopoly responsible for the provision of phone lines) to make much needed investments in relevant technologies. Similarly, a lack of investment into other public utility services continues to hinder the provision of good physical infrastructure for industrial activities, for example electricity. Electricity has been a hindrance to industrial production since the 1970s; a period when power supply was not well-integrated into the construction of large-scale industrial plants.

The survey also shows that a major concern of companies continues to be how to source spare parts and equipment for the day-to-day functioning of the enterprises, or how to maintain production cycles despite the daily occurrence of infrastructure deficiencies.

As a result, capital utilization remains low owing to difficulties related to infrastructure, high (and fluctuating) input and raw material prices, and a lack of access to local and regional markets. The survey also found that local companies still find it very difficult to gain a foothold in local markets, even when they invest in quality products. Respondents considered that the policy regime does not facilitate differentiated product pricing based on quality. Another issue that was repeatedly raised by the interviewees was that local customers often opted to buy foreign goods dumped in the market over locally produced goods.

c. Policy impediments to learning and innovation

Several of the firms interviewed were unaware of the 2011 National STI Policy, and were therefore unable to profit from the policy and incentives contained in this policy in any meaningful manner. Survey respondents also noted that they were unfamiliar with some of the new policy agencies, such as the National Competitiveness Council, which was established in 2013. Other factors identified by surveyed firms as impediments to technological learning and technology-based industrial development are presented in table 3.5 below.

d. Lack of collaborative linkages

Innovation processes rely on the robust interactions of actors both within and outside firms, particularly as these processes act as positive feedback loops to product and process development activities. Survey results show that while firms reported that they are engaged in new products and processes (table 3.3 above), existing weak linkages, low educational qualifications and low levels of government support or investment to promote collaborative partnerships meant that they were not very innovative.

While most firms acknowledged low levels of collaboration as an issue that needed to be resolved, a few companies reported that they were engaged in formal and informal collaborations facilitated by personal networks. Notably, survey results show that informal contacts account for 40 per cent of ongoing interactive collaborations. Survey respondents also cited the poor coordination among agencies competing for relevance as a critical issue.

D. OUTSTANDING ISSUES FOR CONSIDERATION

Nigeria's transformation into an economy of the kind projected in the Nigerian Vision 2020, as well as the Nigerian National STI Policy framework will depend on how and to what extent the country is able to provide an institutional framework that promotes technology-led industrial development. Nigeria's challenges are different from the other two countries (Ethiopia and the United Republic of Tanzania) in this report, in large part because of its dependency on its resource-richness, low levels of technological capabilities, and is faced with a continuing economic boom owing to the rising demand for commodities.[18]

In 2014, fuels accounted for $92 billion of the $98 billion's worth of merchandise goods that were exported by Nigeria. This large share of fuel-based exports estab-

Table 3.5:	Factors preventing enterprises from developing technology and engaging in competition				
	Health care & pharma-ceuticals	Agro-processing	ICT	Other	Overall
Local duties and levies	4.03	4.21	4.09	4.50	4.21
Official corruption	4.19	4.15	4.01	4.25	4.15
Customs procedures and EXIM policy	4.16	3.98	3.76	4.50	4.10
Restrictions in licensing arrangements	4.26	3.92	3.76	4.00	3.99
Regulations, including industrial and innovation policy	4.11	3.92	3.71	3.80	3.88
Patent office delays and other restrictions on testing	4.00	3.96	3.82	3.71	3.87
Others	4.14	3.80	3.63	-	3.86
Municipal regulations	3.92	3.79	3.59	3.93	3.81
Access to land[(1)]	3.65	3.85	3.70	4.00	3.80
Local duties and levies	4.03	4.21	4.09	4.50	4.21

Source: UNCTAD calculations based on Nigeria Field Survey, 2013.
Note: Figures represent the mean of rankings between 1 (no impact) to 5 (prohibitive impact on learning).
[(1)] Registration cost and procedures.

lishes Nigeria as one of the countries with the least diversified merchandise exports.[19] Secondly, despite overall growth, Nigeria's institutional infrastructure is largely unable to promote systemic and sectoral coordination, which is extremely important to promote economic activity of a productive nature.

Nigeria has as a result seen a gradual but significant shift of labour and productive resources away from agriculture and manufacturing to resource-based sectors that have very little or no domestic technological components. This is well-illustrated in figure 3.2, which shows that the composition of Nigeria's exports has only marginally changed over the past two decades. Fuels remained the main source of export revenues and account for more than 90 per cent share in total merchandise exports throughout the period from 1995 to 2014. The share of food and live animals product group fell from about 3 to 2 per cent during the same period, while the share of crude materials and inedible products increased slightly over the past decade.[20]

Merchandise exports have not played a significant role in the acceleration of economic growth during the period from 2000 to 2014. While real annual economic growth rate switched from -0.4 per cent in 1990s to 4.6 per cent between 2000 and 2014, this has had minimal impact on real export growth between these two periods. Revival of economic growth, and thus domestic demand,

on the other hand, led to a surge in Nigerian imports since the beginning of 2000s.

Under these circumstances, promoting development through productivity-enhancing growth is not an easy task despite Nigeria's abundant labour and natural resource endowments (Otsuka, 2012; IDB, 2010).

Box 3.2 below provides the example of Chile, which moved from resource-rich base to a more structurally diversified economy. Chile's move to a more diversified production structure is useful in the context of identifying some policy-based good practices.

E. CONCLUDING REMARKS

As one of the most dynamic countries in Africa, and having recorded the fastest growth rate out of all African countries in 2012, Nigeria is touted to become a new entrant into the club of emerging developing countries in the near future (UNCTAD, 2012). According to the new rebased figures, manufacturing activity is on the rise (at 8 per cent in 2013), and a range of new policy instruments have been put in place to help the economy become more competitive at both the regional and international level.

However, the primary challenge faced by Nigeria is that of moving from planning to implementation, particularly through coordinating its efforts to boost industry perfor-

Figure 3.2: Product composition of Nigerian merchandise exports, 1995-2014 (in per cent)

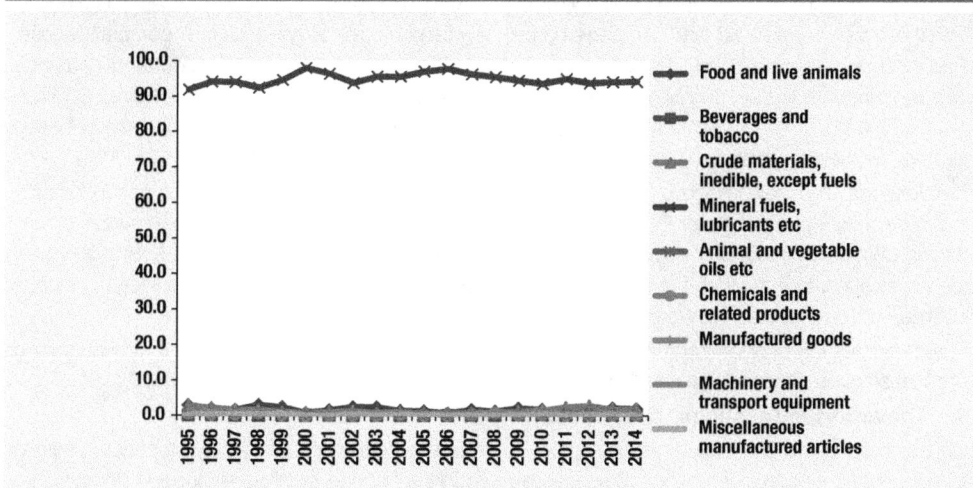

Source: UNCTAD calculations based on UNCTADstat (accessed on 20 October 2015).

Box 3.2:   Chile's industrial and innovation policy mixes: Experience and lessons from another resource rich country

Chile, much like Nigeria, is a resource-rich country, with copper accounting for 60 per cent of exports and 20 per cent of GDP in 2010 (The Economist, 2013). Data show that the real GDP per capita of Chile more than tripled from $ 3,230 in 1970 to $ 9,771 in 2013, making the country one of the best economic performers in Latin America over this period. The country has had its share of successes and challenges over three distinct industrial and innovation policy regimes. The first period covered the state-led import substitution strategy that was pursued prior to 1973; the second, the market-led or economic liberalization policies adopted from 1973 to 1990 following a change in government; and the third, the export-led economic development strategy and reforms which began in the 1990s. Since 2010, Chile has embarked on a long-term economic policy strategy that seeks to consolidate previous economic successes through technological learning and innovation by 2030. Chile's industrial and innovation policy and economic performance and lessons learned are summarized below.

1. State-led import substitution strategy prior to 1973

The industrial policy mix during this period included the protection of infant industries through a high tariff system and import control, and the nationalization of copper mining and private manufacturing firms. The role of the state was not limited to creating an enabling environment for industry; instead, it took on the role of an 'entrepreneur' and became directly involved in mining and manufacturing activities. Whereas direct government involvement under the state-led import substitution strategy failed in several developing countries, Chile's experience delivered mixed outcomes. Data show that the annual real GDP growth rate declined from 9 per cent in 1971 to -5.6 per cent in 1973, and the annual growth rate per capita dropped from 7 per cent to -7.2 per cent over the period between 1971 and 1973.

Key lessons: A key lesson learned is that, the state's direct involvement in economic activities yielded positive results and contributed to increasing the country's annual per capita growth rate to 7 per cent in 1971. The difficulty was that the growth was not sustainable due to fluctuating export prices, falling export earnings and failure to rectify the country's balance of payment problems (Sapelli, 2003). Another lesson is that, the industrial development strategy that was pursued was not coordinated with technological learning and innovation schemes, thereby limiting its positive impact.

2. The economic liberalization policy mixes from 1973 to 1990

Chile launched its liberalization process with an industrial policy strategy involving the reinstitution and return of nationalized firms to the private sector. This led to the withdrawal of the government from its prior direct involvement in manufacturing and production of economic goods. The state's role was limited to creating an enabling business environment for the private sector through the removal of high tariffs, as well as credit and import controls. The state also created supporting institutions, such as ProChile, to promote exports and help exporters to discover niche markets, and the Service de Cooperation Tecnica (SERCOTEC) to provide financial support to SMEs. A very important development at the time was the effort to promote innovation: Fundación Chile was created in 1976 with the mandate to promote technology transfer and innovation and add value to the country's vast natural resources. It is credited with the creation of over 60 new companies using new technologies to enhance their competitiveness. Capacitación y Empleo (SENCE) was also created in 1976 under the Ministry of Labour and Insurance with the mandate to train and offer labour-related services. Real GDP per capita rose from $ 3,114 in 1973 to $ 4,016 in 1990 when the country returned to democratic rule and expanded its market-based strategy with wider application of new technologies and innovation.

Key lessons: The main lesson is that Chile began to emphasize technological learning and innovation in industrial development earlier than other resource-rich countries such as Nigeria. The creation of Fundación Chile in 1976 to promote technology transfer and innovation to add value to the country's natural resources attest to that effort. Fundación Chile was associated with the creation of the salmon industry in Chile, which ranked as the world's second largest exporter in 2006.

Private sector-led economic growth and development became more sustainable with the increasing application of new technologies and innovation.

3. The export-led economic development strategy from 1990 to date

Chile's challenges in diversifying private sector activities became the policy focus in this period. Looking to consolidate its previous economic successes through enhanced institutional support, InnovaChile

Box 3.2: Chile's industrial and innovation policy mixes: Experience and lessons from another resource rich country *(cont.)*

was set up in 2004 with the mandate to subsidize innovation activities of firms and other research institutions. The following year, the National Innovation Council for Competitiveness (CNIC) was established by presidential decree in 2005. CNIC was also given the directive to subsidize innovation activities in the country. CNIC largely focuses on the research institutions and universities (Agosin et al., 2010). InnovaChile provides grants for companies and other research institutions based on project proposals. Besides project grants, and also offers advisory and technical services to companies and other stakeholders in order to foster an innovation and entrepreneurship culture within the country.

Since 2010, Chile also embarked on a global connection programme and provided incentives for a thousand highly innovative entrepreneurs to launch start-ups in Chile. The objective is to provide a learning opportunity for local companies through networking. The government has also sought to encourage local companies to interact and adapt technological best practices from centres of excellence outside Chile. By 2030, the Government of Chile intends to make Chile an innovation hub in Latin America and boost the competitiveness of local companies and the national economy at large (InnovaChile, 2010). The rise in Chile's real GDP per capita from $4,016 in 1990 to $ 9,771 in 2013 can partly be attributed to the policy mixes pursued by the government.

Key lessons: Current government initiatives reflect the state's commitment to promote innovation, entrepreneurship and the competitiveness of Chilean companies in areas beyond traditional technological learning channels, such as FDI and technological acquisition. Employing unconventional innovation schemes, such as 'accelerator and network' programmes for high-tech start-ups in Chile, and also sponsoring local counterparts to tap into the global knowledge bases are cutting-edge initiatives.

Source: UNCTAD.

mance, so that the economic growth can be channeled into sustainable development. A majority of firms, and particularly SMEs, need specific incentives to improve their performance, speed-up access to finance and promote market penetration for their products.

A first step in this direction is coordinating the industrial policy framework with STI policy. This has already been accomplished in theory by identifying the same objectives in the two policy frameworks. The challenge that now remains is to implement this in practice by linking the wide range of policy agencies and incentives to promote technological learning. For this to happen, efforts need to be made to resolve the conflicting objectives of several agencies, set the funding and targets of individual agencies, and articulate the different incentives for the performance of agencies involved in promoting innovation-based industrial performance.

The survey conducted for this chapter, as substantiated by secondary data sources and national reports, shows that Nigeria's institutional infrastructure continues to remain to structurally weak; it is therefore critical to address these limitations in the immediate future. Nigeria has a lot of technology infrastructure, including R&D institutions, dedicated R&D complexes (such as the SHESTCO complex), quality assurance and testing, and technology incubation, some of which are being further strengthened. Despite this, the survey shows that coordination between these agencies needs to be improved, and also that basic infrastructure needs to be strengthened to enable companies to perform more efficiently. However, the country's development objectives are currently compromised by poor coordination among agencies competing for relevance, lack of collaborative linkages, together with the paucity of scientific and technical personnel with the requisite understanding of the S&T system. Clarifying the roles and responsibilities, and increasing budgets of these agencies will help to improve the quality of assistance they render to the private sector.

A comparative perspective of these findings can be found in the concluding chapter of this report (chapter VI).

NOTES

1. Data released from the Nigerian statistics office estimate the value of the country's economy at around $509 billion for 2013, which is nearly twice as high as what was estimated earlier. This outdoes the South African GDP of $315 billion for the same year.

2. The African Economic Outlook 2013 estimates that oil sector contributes to 8.0 per cent of the average annual growth rate of the country, as opposed to the -0.35 per cent of the non-oil sector (AfDB, OECD, UNDP and UNECA, 2013, p. 264).

3. World Bank Database, based on a USD 1 per day poverty estimate. Similarly, unemployment has also been on the rise, going up from 6.4 per cent in 1980 to 21 per cent in 2010 (See AfDB, OECD, UNDP and UNECA, 2013, p. 264; Njoku and Ihugba, 2011, p. 3).

4. UNCTADstat (accessed on 24 September 2015).

5. UNCTADstat (accessed on 24 September 2015).

6. UNCTADstat (accessed on 25 June 2014).

7. ILO LABORSTA database (accessed on 25 June 2014).

8. Ibid.

9. ILO LABORSTA database (accessed on 20 October 2014). These figures differ slightly from those provided by the Nigerian National Bureau of Statistics.

10. For examples of large scale projects that failed in Nigeria, see Poynter (1982), Poynter (1986), Oyelaran-Oyeyinka (1998), among others.

11. Granted by the Central Bank of Nigeria and NEXIM to assist banks to bear the risks in export business and thereby facilitate export financing and export volumes.

12. Granted by the Nigeria Export Promotion Council in 2005 to encourage companies to engage in export business rather than domestic business, especially exported who have exported $ 50,000 worth of semi-manufactured or manufactured products.

13. Granted by the Nigerian Export Promotion Council as a form of export subsidy to compensate exporters of products whose foreign prices had become relatively unattractive due to factors beyond the exporters' control.

14. Granted by the Nigerian Export Promotion Council to encourage exporters to use local raw materials. This has not yet been implemented.

15. In addition to these, there are several incremental milestones listed out in the policy framework for the accomplishment of these goals. For example, it is stipulated that capacity utilization is raised from 54.7 per cent in 2008 to 65 per cent by 2013 (National Planning Commission of Nigeria, 2010).

16. UNCTAD primary survey data on total employment of firms.

17. Many firms were engaged in more than one of these activities that is, production and marketing, or production and product development, hence some of the categories do not add up to 100 per cent.

18. Existing data shows that from 2004 onwards, emerging economies are the main drivers of the resource-boom in African countries, having surpassed the developed world in their demand for natural resources (see UNCTAD, 2012).

19. In 2012, Nigeria scored 0.775 in UNCTAD's export concentration index. Although this score is very high, it is smaller than the country's score in late 1990s and early 2000s. See UNCTAD Stat Database (accessed on 10 July 2014).

20. Exports of food and live animals fluctuated over the course of past two decades in Nigeria's merchandise exports, in part due to changes in global food and commodity prices.

HARNESSING STI POLICY FOR INDUSTRIAL DEVELOPMENT IN THE UNITED REPUBLIC OF TANZANIA

4

CHAPTER IV
HARNESSING STI POLICY FOR INDUSTRIAL DEVELOPMENT IN THE UNITED REPUBLIC OF TANZANIA

a. INTRODUCTION

The Arusha Declaration of 1967 marks the start of state-led industrialization efforts in the post-independent United Republic of Tanzania, and embodies the national ambition that the government set out with: to reduce the country's heavy dependence on agriculture and steer it towards becoming an industry-led economy. Since then, the United Republic of Tanzania has enacted a wide range of industrial development strategies and national plans to promote economic development, job growth and poverty alleviation and continues to strive to promote industrialization. Although there has been stable economic growth between 2000 and 2014, the economy continues to rely heavily on agriculture and low value-added manufacturing. Sustaining future GDP growth rates, therefore, remains a challenge (see figure 4.1).

This chapter is based on a case study of three sectors in the United Republic of Tan-

zania (agro- processing, pharmaceuticals and health care and ICTs). It aims to understand the causes why little structural change has taken place and explain the country's lackluster industrial performance, and the role of policy in the process. The chapter addresses three key questions related to the interface between innovation policy and industrial development. A field survey was conducted for this chapter in collaboration with the Tanzania Commission for Science and Technology (COSTECH) and Tanzania Chamber of Commerce, Industry and Agriculture (TCCIA) to identify policy challenges for innovation capacity and industrial development (see box 4.1).

B. CURRENT DYNAMICS AND STRUCTURAL GAPS IN THE ECONOMY

The challenges faced by the United Republic of Tanzania in promoting industrial growth can be traced back to the 1960s. This sec-

---

Box 4.1: Data sources and field survey in the United Republic of Tanzania

A primary survey using semi-structured questionnaires was conducted by UNCTAD, in collaboration with COSTECH and TCCIA, to understand the underlying drivers of innovation and industrial performance in the United Republic of Tanzania in 2013/2014. COSTECH and TCCIA chose the firms to be surveyed, based on the overall structure of the private sector. Randomized selection techniques were used for the survey.

In addition to the survey, in October 2013 UNCTAD also conducted field interviews and industry visits in the United Republic of Tanzania, in partnership with the COSTECH and the TCCIA, in order to gather information by consulting stakeholders. In addition, a one-day workshop was organized by UNCTAD in collaboration with COSTECH to elicit responses on the key issues in harnessing innovation policies for industrial development in the country, with a wide variety of stakeholders, including national agencies, enterprises and nongovernmental organizations (NGOs).

COSTECH was established in 1986 as a forum that coordinates key scientific and technological institutions. It serves in an advisory role to the government on science and technology (S&T) related issues and their application to bolster socioeconomic development in the United Republic of Tanzania. TCCIA was established in 1988 and currently has 21 regional offices and 90 district centres across the country.

Source: UNCTAD.

tion presents brief sectoral trends, and presents the difficulties faced by the economy from the 1960s up until the present day.

1. Sectoral trends

The United Republic of Tanzania is showing signs of overall economic growth, as depicted in figure 4.1, although this is not attributable directly to an expansion of manufacturing. The services sector is the largest economic activity in the country, accounting for 43.9 per cent of national income in 2013 (table 4.1). This is followed by agriculture, which, accounts for about 33.5 per cent of GDP and 76 per cent of the labour force as of 2013, making it the most important sector in the economy.[1] Viewed in retrospect, data show that the sector's share in national income increased over the past four decades. Despite this, unproductive agriculture remains one of the main challenges in the country's development up until now, along with an over-reliance on extractive sectors and low value-added manufacturing (UNIDO and GURT, 2012).

According to available data, the industrial sector currently contributes the least to GDP when compared to services and agriculture, but has been capturing an ever-increasing share of GDP over the past two and a half decades. This industrial sector growth is mainly accounted for by non-manufacturing sectors, such as mining and construction. Rapid urbanization and increased government investment in public infrastructure have boosted the growth of the construction sector (UNIDO and GURT, 2012). In comparison, the manufacturing sector has witnessed a declining share in industrial activity over the past four decades, similar to the trajectory followed by many other LDCs, but this has become particularly noticeable since the 1980s (see table 4.1). Industrial subsectors, as a result, are also rather weak in terms of employment. Mining and utilities, which jointly account for a total of 5.8 per cent of national income, for example, barely provides 0.7 per cent of the total national employment.

The data presented in table 4.1 has some variations when compared to data recently released by the Tanzania National Bureau of Statistics (NBS), which revised its national accounts statistics in October 2014 by rebasing the series to 2007. According to the new series, the services sector remains the largest with 47.3 per cent share of GDP in 2013, followed by agriculture (31.7 per

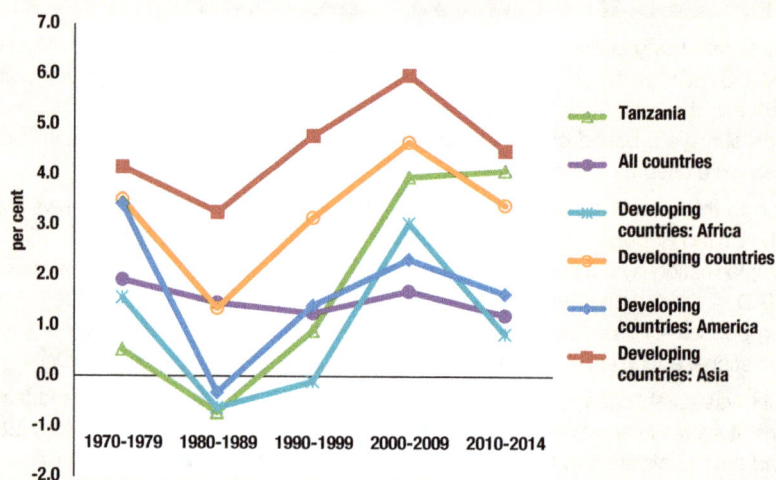

Figure 4.1:   **Real per capita GDP growth rate in the United Republic of Tanzania vis-a-vis other regions of the developing world, 1970-2014 (in per cent)**

Source: *UNCTAD calculations based on UNCTADstat (accessed on 30 October 2015).*
Note: *2014 figures are estimates.*

Table 4.1:	Distribution of the United Republic of Tanzania's GDP by sector, 1970 to 2013 (in per cent)									
Sectors/ Years	1970	1975	1980	1985	1990	1995	2000	2005	2010	2013
Agriculture, hunting, forestry, fishing	17.6	17.7	21.5	27.6	29.8	34.0	31.6	30.0	32.0	33.5
Industry	19.0	17.9	16.6	13.0	16.5	16.8	17.2	20.7	20.8	22.6
Mining & utilities	2.6	1.6	1.6	1.3	1.8	3.9	3.8	4.9	5.9	5.8
Manufacturing	12.0	12.4	11.8	9.3	9.9	8.6	8.1	7.5	7.3	7.2
Construction	4.5	3.9	3.2	2.4	4.8	4.3	5.3	8.3	7.6	9.7
Services	63.4	64.4	61.9	59.4	53.7	49.2	51.2	49.3	47.2	43.9

Source: UNCTADstat (accessed on 24 September 2015).

cent) and industry (21 per cent). The manufacturing sector accounts for 6.9 per cent of national income (NBS Tanzania, 2014).

2. The development of innovation and industrial development policies 1960s until the present day

The United Republic of Tanzania's policy experience in industrial development and STI policies can be broken down into two specific periods. The first distinct period between 1961 and 1980 was characterized by industrial strategies focusing on state-led import substitution. The deployment of World Bank and IMF-sponsored structural adjustment policies during the mid-1980s ushered in a new era in development policies based on export-led and private sector-driven development principles; these continue to form the basis for the country's industrial strategies up until the present day.

a. The first era of industrial development: 1960s to the 1980s

During the early years of independence, the United Republic of Tanzania encouraged investment by private foreign capital (through the 1963 Foreign Investment Protection Act), official development assistance and export expansion (Helleiner, 1976; Wangwe, et al., 2014). During these formative years, the industrial sector remained rudimentary, dominated by foreign-owned companies and largely comprised of agro-processing and low value-added manufacturing (Gray, 2013).

The Arusha Declaration of 1967 marked a significant turning point in the country's de-velopment policies, as it defined a developmental vision for the country and the government's role in achieving more concrete outcomes. Within 24 hours of its adoption, all private commercial banks were nationalized and efforts were made to nationalize all national production initiatives (Nyerere, 1977).

In the period immediately thereafter (i.e. during the second five-year plan, 1969-1974), efforts were made to nationalize existing industries and establish new industrial parastatals. The parastatals and nationalized industries enjoyed subsidies, donor support, heavy state investment and tariff protection, which combined led to expansion and growth in the initial years. However, chronic underutilization of capacity, a steep drop in productivity and mounting state subsidy costs soon began to dampen industrial growth in the 1970s (Gray, 2013).

To address this problem, the government formulated a long-term industrial development strategy (the Basic Industrialization Strategy of 1975-95 [BIS]). The BIS articulated ambitious goals in terms of industrial growth and employment generation, and for the first time recognized to some extent the need for technical and technological inputs in the industrialization process.[2]

By the end of the 1970s, the implementation of BIS was disrupted due to adverse external and internal developments (Wangwe & Rweyemamu, 2004) when the country entered a period of economic crisis. Caused by increased fiscal and trade deficits, spiraling inflation, reduced donor support due to policy disagreements and an acute short-age of imported goods resulting from foreign exchange shortages, the crisis exerted pressure on the government to consider alternative proposals on economic and industrial policies from international agencies during the 1980s.

b. Second era of industrial development: The post-1986 period

In 1986, the United Republic of Tanzania introduced trade liberalization measures through a structural adjustment programme under the auspices of the World Bank and a stand-by agreement with the IMF. As a result of many of these reforms, particularly those related to tariffs and abolition of governmental subsidies, industrial growth continued to plummet during the economic reform and liberalization phase (UNIDO, 2012). Moreover, reforms failed to increase technological capabilities due to limited technological learning, particularly among public enterprises (Wangwe, et al., 2014).

During this period, the strategy for industrial development remained somewhat unclear, as the government was kept busy with emergency structural reforms, mostly to meet conditions set by international agencies. As part of the liberalization package, foreign ownership restrictions in the manufacturing sector were lifted. This had several negative effects: struggling local enterprises could not survive the competition from foreign firms, and the few better performing state-led manufacturing firms were bought up by foreign buyers. By the end of the 1990s, the national enterprise sector had dwindled down to a small percentage of what it used to be, and a large number of local firms had either shut down or were forced to move into the informal sector (see also, ILO, UNIDO and UNDP, 2002).

In order to address these adverse impacts on the industry, a longer-term strategy known as the Sustainable Industrial Development Policy (SIDP) of 1996-2020 was implemented. This policy was based on market-led private sector development, and defined the government's role as that of providing an enabling environment. De-spite its market-led approach, the SIDP envisaged that the government could invest directly in specific industries of critical importance.[3]

In the spirit of this new reform agenda, the Tanzania Development Vision 2025 (TDV) was formulated in 1999. The TDV envisages that the country would become a middle-income country following its transformation into a semi-industrialized economy by 2025. The ambitious targets set out in the TDV are premised on the recognition that the industrial sector will play a central role in the country's new development agenda. However, since a clear policy framework of implementation did not accompany the TDV, there has been a lack of clarity on how these goals would be translated into reality.

Around the same time, the Tanzania Mini-Tiger Plan 2020 (TMTP) was introduced in 2005 to promote diversification of the local economy.[4] The TMTP aimed at replicating the success of Asian tigers by promoting some specific innovation policy measures, e.g. Special Economic Zones (SEZs), to establish an export-oriented manufacturing intensive through, among others, an expansion into technology-intensive sectors.

Despite the government's enthusiasm, the plan failed to garner support from the donor community.[5] Instead, donors convinced the government to adopt the National Strategy for Growth and Poverty Reduction (NSGRP 2005-2010), which focused primarily on poverty reduction (UNIDO, 2012).[6]

The NSGRP emphasized sectoral strategies that include promoting agriculture and resource-based products, as opposed to the earlier vision for a more diversified industrial growth. But at the same time, the five-year plans were reintroduced to promote the TDV 2025 targets with the recognition that there is a need for a clear road map, and this time with some of the earlier emphasis on industrial empowerment. Out of the ten priority investment projects of the first five-year development plan (2011-2016), three projects focus on the industrial sector, namely: (i) Development of

SEZs, especially for electronic goods, farm machinery, integrated textile industry and agro-processing and mineral processing; (ii) Large scale fertilizer production; and (iii) the coal and steel industry. Tanzania is currently in the process of enacting the second development plan (2016-2021).

In addition, the new Integrated Industrial Development Strategy 2025 (IIDS) was enacted in 2011. The IIDS aims to guide the process of *resource-based industrialization* through instruments, such as industrial cluster development, PPPs and SEZs (Government of Tanzania, 2011b). This is somewhat in contrast with studies that have stressed that resource-based industrialization has hindered the growth of the country's productive capacities in manufacturing, and that the focus should not be on resource-based sectors (see for example, UNIDO, 2012, Wangwe, 2013).

3. The evolution of STI policy in the United Republic of Tanzania

The difficulties faced in achieving industrialization despite several industrial development strategies led to a rethink on the role of technological capacity. Particularly, the more recent industrial development strategies were linked by the realization that STI policy is important, and that the lack of technological capacity was responsible for the weak performance of the local industrial sector. These are discussed here at length to trace the linkages in policy articulation and design.

a. Science and technology policy during the pre-1996 period

Early industrial policies in the United Republic of Tanzania focused primarily on state-led industrialization and offered little incentives and attention to certain aspects of STI policies. Noting that the lack of technological focus was impeding the ability of enterprises to process intermediate goods for production efficiently, one of the goals of BIS (1975-1995) was to increase scientific, technical and technological knowledge by expanding the training of industrial workers and to establish centres for industrial services and technology (Kapunda, 2014).

Although the BIS could not be implemented fully, it paved the way for the establishment of important public institutions between 1979 and 1982, including the Tanzania Industrial and Research Organization (TIRDO), the Tanzanian Engineering and Manufacturing Design Organization (TEMDO) and the Centre for Agriculture Mechanization and Rural Technology (CAMARTEC). It also led to the National Science and Technology Policy of 1985, which provided the impetus for establishment of the COSTECH in 1986 (see box 4.1), and laid the foundations for the establishment of a Centre for Development and Transfer of Technology (CDTT) within COSTECH, which became fully operational in 1994. This guided the establishment of the S&T infrastructure until the Sustainable Industrial Development Policy Framework was put in place in 1996.

b. STI policies in the post-1996 period

The country's S&T policy subsequently underwent several changes as a result of the shift to the market-oriented approach to industrialization adopted in the 1990s: the latter emphasized leveraging the private sector for innovation and FDI promotion much more than building indigenous learning. This changed with the Sustainable Industrial Development Policy of 1996-2020, which as discussed in the previous section, aimed to promote 'indigenous entrepreneurial base through orienting the education policy and strategy to emphasize technical education, including strengthening of vocational training institutions and entrepreneurship development' (p. 13-14).

As a result, it provided incentives, such as IPRs and access to credit and recognized the importance of developing, consolidating and strengthening scientific research, technology learning and R&D as contributing elements in the eventual success of industrial sector.[7]

Within the broader policy direction of the SIDP, a revised national S&T policy was enacted in 1996. Although the new policy set the target of increasing the allocation of funds for scientific R&D to 1 per cent of GDP by 2000, half of this target (0.52 per

Box 4.2:	The knowledge infrastructure of the United Republic of Tanzania

The unavailability of skilled labour is one of the country's enduring challenges. While the net enrolment rate in primary education was at 97.6 per cent as of 2008, there was a drastic decline in the same year for secondary education (35 per cent), and tertiary education was a meagre 3.9 per cent.[10]

These declines from primary to tertiary education are drastic, particularly given that the United Republic of Tanzania allocates a significant share of its national income on public education, amounting to 6.2 per cent of its total GDP in 2010. The country's public spending on education exceeds the sums spent on education spending in many other countries worldwide (the world average is 4.9 per cent in 2009). However, it has not translated sufficiently into improvements in education attainment and human capital levels.[11]

This low education attainment levels has not made it possible to train and make available sufficient numbers of R&D personnel. In 2010, the country has about 69 researchers and 16 technicians per million inhabitants – a low number even when compared to other LDCs.[12]

Source: UNCTAD.

cent) was reached as of 2010.[8] The 1996 S&T policy identified specific sectoral objectives and strategies to promote the dissemination of innovation and technology, but also continued the focus on agriculture as a priority sector. Several education and training policies were put in place to improve public access to education and increasing productivity.

The 1996 S&T policy suffered from other shortcomings, *inter alia*:

(i) The sectoral objectives and strategies were not fully translated into policy actions and targets that could be acted upon, as a result of which investments in the country's knowledge infrastructure were not realized (see box 4.2).

(ii) Lack of coordination of the S&T policy with other ministries, particularly the Ministry of Industry and the Ministry for Education, resulted in limiting the impact of the policy on building human skills and strengthening the research system.[9]

Despite the gravity of these issues, it was not possible to address them in a comprehensive way, at least not up until 2005, as the first National Strategy for Growth and Poverty Reduction (NSGPR, 2000-2005) did not manage to adequately integrate STI issues to national development and poverty reduction (Tema & Mlawa, 2009).

But along with the second NSGPR of 2005 and the new industrial development strategy 2025, there has been a fresh impetus on innovation promotion. The current strategy places an emphasis on resource-led industrialization, job creation and poverty eradication. The following changes stand out on targets and policy actions that related to the promotion of innovation:

(i)   STI promotion has been recognized as a core priority. The new Five-Year Development Plan (2011-2016) also spells out certain policy actions to link academic research outputs with productive sectors.

(ii) Industrial cluster development is recognized as the main instrument for promoting industrial innovation in the new industrial strategy.

(iii) The industrial strategy 2025 also recommends upgrading the National Development Corporation (NDC) as an autonomous venture capital fund.

(iv) The 2010 National Research and Development Policy (NRDP) that was enacted to provide an enabling research environment for the promotion of STI emphasizes: innovation and commercialization of research results; the need for priority setting of a national research agenda; and seeks to harmonize the roles of different ministries and institutions that deal with research matters. The NRDP also proposes the creation of a consolidated National Research Fund.

c. Current STI context

Presently, COSTECH is a central actor to promote STI in the country, acting under the Ministry for Communication, Science, Technology and Innovation. In addition, it also houses the centre for technology transfer (CDTT). The Government of the United Republic of Tanzania also introduced certain financial measures, such as fee waiver for R&D activities, tax incentives for properties of R&D facilities, licensing of technology and innovation and revised the Finance Act in order to incentivize innovation, particularly in low-cost technologies, products and services.

The National Planning Commission has a mandate to advise the President on development planning policy and strategy also plays a major role in the implementation of innovation-related policies. However, in order to promote the coordination of STI efforts and to revisit national priorities in the light of the Vision 2025 document and the IIDS 2025, a revised National STI Framework is currently being prepared and pending Cabinet approval. This revised policy framework is being prepared in conjunction with an ongoing reform of the national system of innovation.

The draft proposed policy framework could broaden the definition of COSTECH by adding 'innovation' to the agency's title and mandate. If the changes are adopted, the new Commission would be called Tanzania Commission for Science, Technology and Innovation (TCSTI), and would become the government's principal advisory organ on STI issues.

The draft STI Act proposes the establishment of a Centre for Innovation and Technology Transfer and a National Fund for the Advancement of STI (or a national fund on innovation) to provide loans or grants to research activities, including innovation

projects. Although there have been discussions on such a fund, its structure, management and agency affiliation have yet to be decided.

C. INNOVATION CAPACITY AND INDUSTRIAL DEVELOPMENT: RESULTS OF THE FIELD SURVEY

Two important aspects of the country's industry are of relevance to understand the survey results. The first aspect relates to the nature of its private sector, which has long been noted for its polarized structure, with only a small number of firms that have an export-orientation and a large informal sector producing low-value added products for the domestic market with little productivity (ILO, UNIDO and UNDP, 2002). More recent figures show that the informal sector has expanded and now accounts for about 48 per cent of the economy (see Osoro, 2009; ESAURP, 2012).

According to estimates, about 88 per cent of the country's private sector firms are classified as micro-enterprises with less than 5 workers, contributing to an estimated one-third of GDP.[13] A recent survey pegs the total number of MSMEs at over 3 million (Ministry of Trade and Industry, 2012). The same study also suggests that a minority of these enterprises are of medium size, given that only 3.9 per cent of the businesses are formally registered under the Business Registration and Licensing Agency. In sum, the bulk of the economy is comprised of informal, micro- and small-sized firms with a few medium and large companies.

The sectoral spread of the enterprises is also uneven, given that a majority of MSMEs (85.5 per cent) operate in services, while 0.4 per cent and 13.6 per cent are active in agriculture and manufacturing, respectively (Ministry of Trade and Industry, 2012). In manufacturing, 91 per cent of the manufacturing firms are privately owned, and 97 per cent of total number of manufacturing firms employ fewer than ten employees (UNIDO and GURT, 2012).[14]

These broad patterns were also captured by the survey, which covered a total of 144 firms, 50 of which belonged to the health care and pharmaceuticals sector, 54 to the agro-processing sector and 28 to the ICTs sector. Of the 114 companies that provided employment figures, 60 firms in the entire sample can be classified as micro-enterprises (employing less than 10 employees), 38 are small-scale enterprises and 16 were medium- and large-scale companies, of which seven were in the pharmaceuticals and health care sector.

The second aspect of relevance to understand the survey results is that at least two of these three sectors have experienced significant growth rates over the past decade in the national context.[15] This was particularly striking in the agro-processing sectors, which registered a cumulative increase of 358 per cent over the period between 1985 and 2012.[16]

1. Sector snapshots

The agro-processing sector in the United Republic of Tanzania is built on local agricultural produce, such as processed cashews, coffee and dairy products. Over time, it has expanded to include not only traditional agricultural varieties, but also cash crops, such as cotton, coffee, cashew nuts and pyrethrum (Oyelaran-Oyeyinka and Gehl Sampath, 2007). It is fairly low-scale, with a large number of companies employing less than five employees, or also often simply in-house operations run by women or entrepreneurs relying on local, often self-sourced financing. They mostly undertake self-learning to acquire skills, and show a strong entrepreneurial culture of high achievement under conditions of limited opportunities. The only exceptions to these are some large-scale enterprises with sophisticated technological machinery used for processing. In the dairy business, there are few large milk producers and processors. Little cooperation exists between SMEs, except through cooperatives.

The pharmaceutical sector contains key companies engaged in the production of certain important pharmaceutical products. A handful of pharmaceutical companies produce about 30 per cent of all over-the-counter and prescription medicines. The survey therefore also covered several small- and medium-sized companies producing herbal food and traditional medicines based on local traditional medicinal knowledge.

In the ICTs sector, there are no hardware producers in the country and it was difficult to identify firms that are active in software development. The sample therefore included ICT product accessories suppliers, local Internet companies and distribution companies.

2. Survey results: Innovation opportunities and performance

As detailed in chapter I, the survey questionnaire sought to understand the underlying nature of innovative activity by capturing many aspects of the way firms' perceived their products, including whether they thought it as new to the firm itself, the local market, the region or the world. It then sought to ascertain the source of these technologies, and the basis for technological upgrading.

a. Nature of innovation in the three sectors

A large number of the companies surveyed were engaged in distribution, marketing and supplying, but about half of those active in pharmaceuticals and ICTs reported having engaged in new process and product development-related activities (table 4.2). However, most of the respondents noted that their products are new either to the local firm or to the local market, which points to the incremental nature of the process and product innovation patterns.

On the whole, less than a quarter of the firms in the three sectors were engaged in innovation activities that could result in products or processes that are new outside Tanzania's national market, and this percentage is even less in the case of the

Table 4.2: Distribution of firms carrying out new product and process developments				
	Health care and Pharmaceuticals	Agro-processing	ICT	Overall
Number of Firms[1]	50	66	28	144
Product Development[2]	12/19	19/39	9/15	40/73
Share (%)	63.2	48.7	60.0	54.8
Process Development[3]	10/19	14/32	7/14	31/65
Share (%)	52.6	43.8	50.0	47.7
New products are new to (per cent):				
Firm	48.0	68.3	47.4	57.6
Local market	52.0	36.6	47.4	43.5
Regional market	12.0	19.5	26.3	18.8
Global market	12.0	7.3	0.0	7.1

Source: UNCTAD calculations based on Tanzania Field Survey, 2014.
Note: Figure may add up to more than 100 per cent as new product could be new to the firm, local mar-
ket, regional market or global market at the same time.
Figures only include firms that reported new product development.
[1] Number of firms that participated in the survey.
[2] Number of firms that carried out product development out of the total number of firms responding to this question.
[3] Number of firms that carried out process development out of the total number of firms responding to this question

global market. In pharmaceuticals and health care, about 12 per cent of the firms reported that they were engaged in developing products that are new to the world. But the interviews covering these companies showed that many of these products included traditional medicine, herbal food and pharmaceutical drugs.

b. Sources of technological information

The United Republic of Tanzania has an extensive S&T infrastructure for R&D in most sectors of the economy. Current statistics show that the R&D system consists of 62 research institutes covering agriculture, livestock and forestry (28), industry (4), medical (11), wildlife and fisheries, as well as some private sector research institutions.

Many of these public sector research institutions are tasked with providing extension services for agriculture, which are important for firms engaging in agro-processing. The country also has several technology incubation centres focusing on different aspects of relevance to the three sectors surveyed. For example, COSTECH hosts the Dar es Salaam Incubation Centre established through a partnership with InfoDEV on Information and Technology. The Engineer-

ing and Manufacturing Design Organization has a similar partnership focused on agribusiness and the Small Industries Development Organization has mixed incubators in several regions, including in Dar e Salaam.

However, the survey found that two sources of innovation, namely adaptation and collaboration with industry associations, were important activities at the firm-level, underscoring the importance of *incremental, adaptation-driven innovation* processes within firms. Support from intermediary organizations (such as the technology incubation centres) was quoted as the third important source, above technological know-how transfer, which was ranked as the fourth source.

c. Technological intensity of firm-level activities

In order to understand the constraints that account for the low-technology intensity of firms in the country, survey respondents were asked to rank the contributions of various sources to firm-level innovation activity (see table 4.3). The figures presented in this table represent the average rankings of these factors by surveyed firms (ranging from 1 if it was least important to 5 as most important). Hence, any mean ranking above 3 implies that the source was rated as highly important, or below 2 implies that it was rated as not so important.

Most firms surveyed and interviewed acknowledged that the lack of human skills was debilitating and rated the contribution of scientific and skilled manpower to be highly important, with a mean ranking of 3.53. This was followed by the quality of local infrastructure as the second most important factor influencing new product/process development, followed by funding constraints and government incentives. The firms surveyed also noted the relevance of IPR protection, which often hinders the availability of knowledge. This, coupled with the fact that technology transfer initiatives

are not well-supported in the current policy framework, was cited as a major problem by several interviewees. The interviewed firms also felt that local R&D institutes and universities were not very helpful to building technological capacity, since they were engaged in basic research and teaching, as opposed to applied research that could have direct bearing on firm-level activities. Firms also often noted that public sector research results should be made more relevant to industry.

As can also be seen in table 4.3, intersectoral variations are captured. For example, the ICTs sector is more technologically intensive as firms require some level of technological capabilities to survive. This

Table 4.3:   Contribution of various sources to new product or process development

	Health care and pharma-ceuticals	Agro pro-cessing	ICT	Overall
Scientific/skilled manpower	3.17	3.49	4.16	3.53
Qualify of local infrastructure services	3.62	3.59	3.08	3.52
Availability of venture capital	3.15	3.14	3.00	3.12
Government incentives for innovation	2.85	3.27	2.93	3.09
Intellectual property protection	3.04	2.97	3.25	3.04
Local R&D institutes for R&D collaboration	3.17	2.69	2.67	2.85
Local universities for R&D collaboration	2.69	2.85	2.86	2.80
Participation in government-firm technology transfer coordination councils	2.77	2.50	3.45	2.75
Participation in local SMI development schemes	2.63	2.76	2.67	2.70
Transfer of personnel to local firms or R&D institutions for training	2.48	2.59	3.23	2.66

Source: UNCTAD calculations based on field survey, 2014.
Note: Figures represent the mean of rankings between 1 (not important) to 5 (extremely important).

Box 4.3:   Benefiting from foreign partnerships: The case of Claphijo

Claphijo represents a good example of South-South cooperation at the SME level. It is a family owned company that has since 2002 produced dried fruits and vegetables, such as dried mango, pineapple and banana. The method of sun drying as opposed to using electricity-operated machines is a more cost-effective method. With the help of some cooperation initiatives, the company developed its drying facility in four phases: It first received support in the form of machinery and know-how from a German partner, then the Department of Chemical and Process Engineering of the University of Dar-es-Salaam, followed by the Tanzania Traditional Energy Development Organization (TATEDO). It also received support from the Government of the Plurinational State of Bolivia. Claphijo, which was initially producing 200 kilograms of dried fruits and vegetables per year, started manufacturing 5-7 tons after the third expansion phase. The fourth phase is expected to include the installation of an electric dryer in the facilities.

Source: UNCTAD.

explains the higher ranking (all factors are rated above 2.5, but some factors, such as skilled manpower is rated as extremely important for new product/ process development) in this sector.

There are, however, some successful cases where companies have managed to rely on collaboration in order to succeed (see box 4.3).

3. Survey results: Sectoral weaknesses, innovation constraints and industry performance

The survey and field interviews found that a large number of the local companies often operated on the fringes of the local economy and were struggling to technologically upgrade and remain competitive. The survey responses also indicated that the country's innovation system is still quite fragmented, as discussed at length below.

a. Innovation constraints

A first innovation constraint is at the formalization stages itself. While several studies have found a direct link between formalization of firms and overall economic growth, Tanzanian firms often find it very hard to navigate the regulatory frameworks to register themselves. Secondly, there are a large number of incentives for industrial development that often do not function well and are difficult to make use of. Hence, the survey and interviews showed that in the absence of clear assistance, grants, subsidies or other such support structures, firms were not often prepared or willing to undertake the difficult process of registration, least it led to a lengthy process with multiple fees, affecting profitability further. A third set of issues raised by interviewees was that the innovation framework was not very well coordinated and did not facilitate interactive learning and collaboration. This has resulted in local firms not interacting beneficially with universities, public and private research institutes and other intermediate organizations.

i.   Arduous regulatory frameworks

In all the three sectors studied, firms reported difficulties in regulatory environments, partly due to the inability of policymakers to foresee and target technological growth in industrial development. These were related to extremely stringent regulations or a lack of regulations in key areas. An example of stringent regulation that could hinder learning is the country's data protection regime in the pharmaceutical sector. The rapidly changing ICT sector is an example where there are often grey areas in terms of regulation that inhibit entrepreneurship and ability to perform, such as spectrum allocation regulation, including regulation for licensing, which is needed to benefit from TV white spaces for Internet access.

Indeed, survey results indicate the high relevance of excessive and restrictive regulations on the firms in all three sectors (see table 4.4).[17] Licensing restrictions and regulations at the municipal level are the second and third most important factors impacting on firm-level activities. According to the survey, other regulatory practices, such as Customs procedures and EXIM regulations were also rated as being detrimental to promoting innovation in the current context.

ii. Technology transfer and technology incubation issues

There is a pressing need to pursue technology transfer systematically through existing and new national venues. Survey respondents stressed the need to have better technology transfer and technology licensing information and support services. They ranked the issue of promoting enabling rules and agencies to promote technology transfer as highly important (4.14) in efforts to support the introduction of innovation efforts in their companies (table 4.5). Given the large share of MSMEs, companies in this category were worst affected by the absence of services that could help them engage in routine technological upgrading. The survey results also show that addressing loopholes in current patent system and speeding up the patent applications pro-

Table 4.4: Factors preventing Tanzanian enterprises from developing technology and becoming competitive

	Health care and pharmaceuticals	Agro processing	ICT	Overall
Local duties and levies	3.54	3.57	3.38	3.52
Restrictions in licensing arrangements	3.41	3.41	3.76	3.51
Municipal regulations	3.48	3.59	3.39	3.51
Access to finance	3.18	3.95	3.15	3.49
Access to land[1]	3.32	3.62	3.00	3.42
Customs procedures and EXIM poliy	3.41	3.52	3.13	3.39
Official corruption	3.27	3.30	3.08	3.25
Regulations	3.27	3.25	3.00	3.23
Patent office delays and restrictions on animal testing	3.22	2.60	2.86	2.86
Transfer of personnel to local firms or R&D institutions for training	2.48	2.59	3.23	2.66

Source: UNCTAD calculations based on Tanzania Field Survey, 2014.
Note: Figures represent the mean of rankings between 1 (not important) to 5 (extremely important).
[1] Registration cost and procedures.

Table 4.5: Areas where government or other institution's support is critical to devise new innovation strategies

	Health care and pharmaceuticals	Agro processing	ICT	Overall
Improve Customs procedures and EXIM policy	4.21	4.19	4.15	4.19
Enable rules and agencies to promote technology transfer	4.23	4.00	4.25	4.14
Deal with loopholes in the Patent Amendment Bill	4.30	3.80	4.13	4.04
Create a more enabling R&D environment	3.96	3.87	4.18	3.96
Access to land[1]	3.91	3.97	3.90	3.94
Improve speed of processing patent application	3.65	3.70	4.20	3.77

Source: UNCTAD calculations based on field survey, 2014.
Note: Figures represent the mean of rankings between 1 (weak effect) to 5 (very strong effect).
[1] Registration cost and procedures.

cedures were other important areas where government can make a difference (with a rating of 4.04 and 3.77, respectively).

iii. Local business practices and support to SMEs

Given the uncertain innovation and industrial environment in which they operate, local businesses are used to having a short-term focus on how to survive and sell their products. There is a severe lack of support to smaller firms, which impedes their ability to perform. There is not only a need to assist in providing/initiating good business models for cooperation, but also to help in improving the negotiating position of local entrepreneurs in their dealings with multinational companies (MNCs) operating within the United Republic of Tanzania.

Lax implementation of the policy frameworks both on industrial policy and STI, particularly, with respect to processing governmental schemes to support industry needs to be addressed. For example, in incubator projects, the share of returns allocated to a local incubator may be too low.

To underscore this point, table 4.6 contains a summary of how many firms surveyed participated in government programmes or received governmental assistance. The figure remains below 40 per cent for each of

Table 4.6:	Share of firms participated in government programs or received government assistance during the last five years (in per cent)			
	Health care and pharmaceuticals	Agro-processing	ICT	Overall
Participating in government sponsored R&D programmes	36.0	30.6	27.8	31.6
Received government assistance for R&D	34.3	28.3	15.0	27.7

Source: UNCTAD calculations based on field survey, 2014.

the three sectors, and in some instances, e.g. the ICT sector, less than 20 per cent of the firms reported having received governmental assistance.

Survey respondents also rated corruption as an important cause of low-technology development and innovation in the country (see table 4.4). Most companies interviewed during the field survey also noted that access or personal connections to policymakers is often useful to make headway.

iv. Finance

Access to finance for entrepreneurs is a key constraint for firms in all three sectors. It is among the main factors preventing companies from engaging in technological development and innovation (table 4.4).[18] There are also few local banks and a shortage of sophisticated financial products/ schemes for innovation. For example, there are four Kenyan banks in the country, but no large local bank to offer credits to start-ups and small enterprises. Most firms suffer from not having sufficient start-up financing, and reported to be often unable to operate within a medium-term or longer-term timescale without financial security (interviews). There

are, however, some private sector initiatives to facilitate pro-poor access to finance in the market by using new ICTs (see box 4.4).

Local tax structures, survey respondents noted, are also an impediment and need to be made more conducive for effective functioning of companies (table 4.4). The survey interviews showed that local companies can now benefit from loopholes in tax regulation to perpetuate their tax-exempt status, for example by changing their international partners every few years. The tax climate also seems to account for a large amount to promote and support a culture of innovation and entrepreneurship is therefore key to future success.

v. Low/ expensive access to intermediate inputs of production

Another relevant factor to low competitiveness raised by several companies are the heavy duties paid on intermediate inputs required for production.[19] For example, an important issue for the Tanzanian dairy sector in general is the cost of packaging. Bottles are bought from Kenya because they are not available locally, and transported by road to Dar-es-Salaam. In addition to the

Box 4.4:   Pro-poor access to finance: Maxcom Africa

Maxcom is an ICT company that integrates mobile money operators, such as M-PESA, to facilitate point-of-sale payments for customers that do not have bank accounts. Its services consist of integrating mobile money operators, integrating bank services and providing payment gateways. It is among the companies that operate under COSTECH's Dar Teknohama Business Incubator. The company has signed a Memorandum of Understanding with the Ministry of Technology, Science and Higher Education to facilitate the payment of bills and distribution of mobile money. Maxcom started providing these services in early 2010 and it is now operating in the United Republic of Tanzania, Rwanda and Burundi. Maxcom offers its 'MaxMalipo' services through a network of 6,500 agents located across all regions of the United Republic of Tanzania.

Maxcom fills a vacuum in pro-poor financial resources and public revenue collection. The fact that the government has supported Maxcom's business model has played an important role in its success.

Source: Field interviews, UNCTAD.

high price for the imported bottles, Customs levies are costly and customs formalities time consuming. These constraints often hinders companies from expanding production into related sectors, for example, some of the diary companies surveyed said they were unable to engage in the production of butter because they were unable to procure reasonably good packaging materials locally.

b. Key issues arising in policy coordination between industrial and innovation policy

The survey and field interviews shed light on the numerous opportunities and the difficulties to harness innovation towards industrial development. A first set of issues relate to the difficulties within the STI policy framework itself. The second deals with establishing smooth coordination between STI policy and the country's industrial development strategy, as discussed below.

i. Fragmented policy support apparatus

Survey and interviews found that inadequate coordination in the formulation and implementation of the innovation-related policies among relevant actors had drastic consequences. Ministries often tend to ignore policies from other ministries in the development of their own policies, despite the underlying principle of inclusivity. This results in the absence of linkages between policies, leading to difficulties in efforts to harmonize the implementation of these different policies.

The National Strategy for Growth and Poverty Reduction and the Industrial Development Strategy 2025 both have laid out the broad contours of STI policy. The Ministry of Communication, Science and Technology is responsible for policy formulation on scientific research and technology development. The Ministry of Industry has the overall responsibility for the IIDS 2025, but is expected to operate in collaboration with other ministries involved in the national Poverty Reduction Strategy, e.g. the Ministries for Trade, Education, Agriculture, Health, Environment and Labour. However, it is often not clear which Ministries deals with which issues and how collaboration can be established.

The National Planning Commission (NPC), which has responsibility for a wide range of policy areas and a mandate to advise the President on development planning policy and strategy, also plays a major role in the implementation of STI-related policies. The NPC is currently formulating the second five-year plan, with the intention of addressing some of these linkages, particularly those between innovation policy and industrial development.

ii. Overlapping policy measures and incentives

In line with their respective mandates, governmental agencies operating within these various ministries bear the responsibility for implementation of these policies. However, as survey results show, there still remain a large number of industry measures and incentives that the industry could better avail of, if these are implemented and coordinated effectively. For example, the heavy costs and levies on imports of intermediaries, the lack of business models to partner with larger MNCs operating within the country and heavy regulatory restrictions for compliance in various industries are issues that Tanzanian companies have to face on a daily basis. Addressing these will help the local industry regain profitability.

iii. Issues within the STI system

What has clearly stood out in the survey was that, as of the early 2000s, there has been a gradual shift in the way the Government of the United Republic of Tanzania has approached STI policies; this is particularly evident from the gradual but steady focus on innovation issues. However, despite some improvements, inter-linkages between R&D activities in the public sector and the private sector remain weak.

In the current context, it remains imperative to identify existing initiatives and determine how these can be strengthened to improve weak systemic coordination among various institutions and other actors responsible for supporting innovation in the country. Moreover, a shortage of funding for agencies tasked with STI-related activities, particularly COSTECH activities, has also undermined the performance of the country's institutional structure for S&T, and this need to be further addressed.

There has also been a proliferation of initiatives without much thought paid to consistency or effectiveness. For example, the NRDP of 2010 calls for a research fund to be set up. This however, has not been done. The draft proposed STI Act also includes a National Innovation Fund. It therefore remains an urgent imperative to clarify how and in which ways each of the initiatives should work and how they will be coordinated and monitored.

E. CONCLUDING REMARKS

STI and industrial development frameworks in the United Republic of Tanzania have not evolved in an entirely synergistic manner. Therefore, although there are a lot of positive developments in the current context, several institutional factors continue to impede coordination. Many of these difficulties can be traced back to the several changes that the country's industrial policy has undergone over the past decades, and the way in which the STI policies have evolved alongside.

In the present context, the following need to be urgently addressed. Firstly, the country's industrial policy framework emphasizes resource-based industrialization, but there is no clarity on what this means, and how it can be achieved. Furthermore, it stands in stark contrast to reality, wherein resource-dependence has dampened the development of productive capacities in many manufacturing industries and prevented a suitable diversification of the country's industrial base. A clear elaboration of this strategy is an important next step, as is the need to clearly link it to the upcoming five-year plan.

Secondly, the country's industrial development policies, despite some improvements,

have failed to improve its physical and knowledge infrastructure, which urgently need to be addressed.

Thirdly, a lack of focus on innovation capacity and linkages and a lack of systematic approaches even within the national S&T policy have hampered the country's capacity for technology adoption and innovation, thereby affecting its international competitiveness and growth rates. The focus on innovation should therefore be strengthened in the new STI policy, especially with a focus on SMEs.

Fourthly, gaps and lack of connectedness among existing industrial development plans, sectoral strategies and the national S&T policy, coupled with a road map on how these policies ought to be coordinated has made it difficult to ensure developmental outcomes. New efforts in the upcoming five-year plan should be aimed at clarifying this, setting out a clear road map linking industrial expansion with the new STI policy, outlining policy processes that can facilitate this.

While addressing these issues, the following aspects call for attention.

(i)   The economic development of the United Republic of Tanzania relies significantly on MSMEs, including in the informal sector. Both the STI policy and the industrial development strategy need to bear in mind the needs of these firms if targets set in policy documents or strategies are to be realistic and attainable. Currently, within the policy context of the IIDS 2025 and the NSGPR, there is a significant potential to strengthen innovation at a macro, industrial level since they aim to improve the overall business climate and provide overarching incentives. Addressing the weak support provided to SMEs or micro-enterprises in the two strategies is also needed.

(ii)   SMEs require additional policy support for technological learning, innovation capacity and overall competences to master and excel in business practices. Within the current policy framework, there is no significant attempt to strengthen the link between local research and innovation and the needs of the local firms. For example, the implementation of the NRDP has not been very successful because the current developmental plan has no budget allocation for most of the proposed policy actions in NRDP. The Industrial Strategy 2025 also lacks a comprehensive national framework for technology diffusion and knowledge transfer to local industries.

(iii) SMEs also need more support in another direction namely, by improving the regulatory framework. The government needs to overhaul its regulatory system and simplify its procedures in order to promote local production. Particularly, the government should revise its existing practices so as to encourage entrepreneurship in its technology intensive sectors.

(iv) The lack of donor support to development programmes has impeded implementation of the country's industrial development plans and STI policies. Donor support programmes account for a major portion of the government's budget but it comes with certain conditions, including prioritizing interventions in social sectors and improving public governance (UNIDO, 2012). Excess emphasis on policies designed to limit the role of government in the economy has been hindering development impact of aid programmes. In terms of donor funding, therefore, the government needs to establish a partnership that focuses on the country's specific national priorities.

(v) Although there are several laudable targets in the policies for industrial development and the Tanzanian Vision, these are not matched with adequate implementation strategies and funding for the institutional operationalization of these targets. Monitoring of policy effectiveness is also particularly weak. Ensuring the success of the current five-year plans for sustainable industrial development rests on coherent policy coordination and implementation, especially of the kind that is based on the industry needs for innovation-led growth.

NOTES

1. UNCTADstat. ILO data ends in 2006 for agricultural employment, where it accounted for 75 per cent of the total share.

2. BIS promoted the establishment of medium and small-scale industries in the regions, districts and villages, which lead to establishment of Small industry Development Organization (SIDO).

3. Government of Tanzania (1996). Sustainable Industrial Development Policy – SIDP 1996-2020. Dar e Salaam, Ministry of Industries and Trade of Tanzania.

4. Japan Development Institute (http://www.jditokyo.com/en/projects-3.html)

5. In 2005, 20 per cent of the budgeted public expenditure was financed through donor support and 80 per cent of the total amount spent on development was funded through donor projects (Lawson, Booth, Msuya, Wangwe, & Williamson, 2005).

6. The government also updated the Sustainable Industrial Development Policy (SIDP) 1996-2020 to match the new development framework outlined in the FYDP and LTPP.

7. Government of the United Republic of Tanzania (1996). The National Science and Technology Policy for Tanzania. Dar es Salaam, Ministry of Science, Technology and Higher Education.

8. World Bank's WDI database; 2010 is the most recent year for which data is available.

9. The Government of Tanzania 2010 National Research and Development Policy, as prepared by Ministry of Communication, Science and Technology.

10. World Bank WDI Database (accessed on 22 July 2014)

11. Ibid.

12. UNESCO Database (accessed on 22 July 2014)

13. Data as per the note prepared by TCCIA and COSTECH for UNCTAD, 2014 (on file with UNCTAD).

14. UNIDO and the Government of the United Republic of Tanzania (2012).

15. UNCTAD calculations based on NBS Quarterly Production of Industrial Commodities: 2004-2012 (National Bureau of Statistics- United Republic of Tanzania, 2013).

16. UNCTAD calculations based on NBS Quarterly Production of Industrial Commodities: 2004-2012 (National Bureau of Statistics- United Republic of Tanzania, 2013).

17. The survey conducted by Wangwe, et al. (2014) also finds that manufacturers consider the multiplicity of taxes at both local and national level as impediments on their development along with excessive regulations.

18. According to Wangwe, et al. (2014), high credit interest rates and elaborate credit procedures are among the main challenges inhibiting the development of Tanzanian firms.

19. Wangwe, et al. (2014) estimates that firms import about 70 per cent of their inputs from abroad.

PROMOTING INNOVATION POLICIES FOR INDUSTRIAL DEVELOPMENT IN ETHIOPIA

5

CHAPTER V
PROMOTING INNOVATION POLICIES FOR INDUSTRIAL DEVELOPMENT IN ETHIOPIA

A. INTRODUCTION

Ethiopia's Growth and Transformation Plan (GTP) of 2010-2015 (hereafter referred to as GTP I) is an ambitious policy document aimed at creating an enabling business environment and nurturing the growth of the country's industrial sector (MoFED Ethiopia, 2010). These goals are a follow-up of Ethiopia's developmental plans over the past two decades, where the country has heavily invested in infrastructural development, expansion of exports and increased expenditure to enhance pro-poor growth (AfDB, 2010). In order to ensure continuity, the GTP I envisages a range of policy measures and incentives, including tax holidays, duty-free capital goods imports and the creation of industrial cluster zones. It also sets out several targets for productivity growth, capacity utilization in industry and exports earnings across various sectors, and seeks to link economic growth with poverty reduction and other development targets. Ethiopia was expected to invest over $75 billion to implement GTP I, which it hoped would achieve double digit (11-15 per cent) annual GDP growth between 2010 and 2015 (MoFED Ethiopia, 2010).

The GTP II (2015-2020) is expected to continue these efforts.

Ethiopia has previously undergone several important policy shifts in its industrial growth and transformation. These shifts have often been rather radical and have altered the nature of institutional support to industry. Starting out with a policy emphasis on building capacity in the private sector in the 1960s, Ethiopia moved to a state-led development strategy in the period between 1974 and 1991.

Figure 5.1: **Real per capita GDP growth rate in Ethiopia vis-a-vis other regions of the developing world, 1970-2014 (in per cent)**

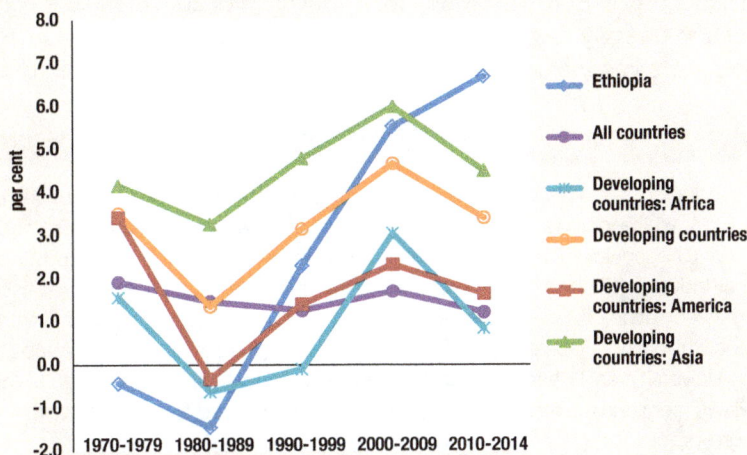

Source: UNCTAD calculations based on UNCTADstat (accessed on 20 October 2015).
Note: 2014 figures are estimates. Ethiopia's 1990-1999 growth average covers the period between 1992 and 1999 due to break in the data in 1991.

The Ethiopian government began implementing industrial development policies, which emphasized private sector-led industrial growth and exports promotion after 1991 (Gebreeyesus 2013). The objective of these policies was to increase export-earnings and create a well-diversified industrial-base. On the positive side, growth rates have improved drastically, from a real GDP growth rate of -6.7 per cent in 1991 to 13.6 per cent by 2004 and 7.6 per cent in 2014 (see figure 5.1).

Ethiopia subsequently launched important policy initiatives to promote industrial and socioeconomic development to consolidate its positive economic performance of the 1990s. Among these policy initiatives is a new national STI policy,[1] which is being implemented alongside GTP I to facilitate the emergence of innovation capacity.

Despite these positive trends, ensuring productivity-enhancing growth still remains a fundamental challenge, and although Ethiopia experienced a sustained increase in GDP growth rates, this was not accompanied by structural change (see table 5.1). Ethiopia's export structure is also highly concentrated around coffee exports, which account for over a third of its export revenues and also poses some risks to overall diversification into other sectoral activities.

This chapter seeks to analyse the role that innovation could play in Ethiopia's efforts to industrialize and diversify its economy, and is informed by a primary study conducted by UNCTAD in Ethiopia, which included a questionnaire survey similar to what was conducted in the other two countries presented in this report, but only of two sectors: the agro-processing and pharmaceutical industry (see also box 5.1).

B. REVIEW OF INDUSTRIAL DEVELOPMENT AND INNOVATION CAPACITY IN ETHIOPIA

Industrial policy interventions aimed at building the Ethiopia's industrial base have been underway since 1960s, but it was not until the 1990s that Ethiopia began to record impressive economic progress. The economic reforms of the 1990s are credited with the industrial progress and sustained real growth rates seen in Ethiopia (see figure 5.2).

For example, while the average real GDP growth rate initially declined from 2.0 per cent in 1970/80 to 1.4 per cent in 1981/91, it increased to 10.7 per cent in 2003/14 (see figure 5.2).[2] At the same time the average real GDP per capita growth rate declined from -0.2 per cent in 1970-1980 to -1.8 per cent in 1981/91, but increased to 2 per cent in 1992/2002 and reached 7.8 per cent in 2003/14.

As in the case of much of Africa, the financial crisis of 2007/08 had a negative impact on Ethiopia's annual growth (see UNCTAD, 2009 and 2010). Real GDP fell from 13.6 per

---

Box 5.1:   Additional information on the field survey in Ethiopia

UNCTAD conducted a primary data survey of enterprises and stakeholder institutions in Ethiopia in 2013 and 2014, which included on-site visits and interviews with 44 stakeholders, including 29 enterprises and 15 agencies. The agencies interviewed in this field survey included: the Directorate of Chemical Industry, Ministry of Industry; Agricultural Transformation Agency; Ethiopian Coffee Processing Association; Food, Medicine and Health Care Administration and Control Authority; Addis Ababa University - Research and Technology Transfer Directorate; Addis Ababa Science and Technology University; and Pharmaceuticals Supply and Fund Agency. In addition to these face-to-face interviews, semi-structured questionnaires were administered to randomly chosen enterprises active in agro-processing (15 firms) and pharmaceutical (14 firms).

In addition to survey findings and face-to-face interviews, country reports and documents, articles and other archived accounts of Ethiopia's development have informed the analysis.

Source: UNCTAD.

Figure 5.2:   Trends in average real GDP growth rate and GDP per capita growth rate, 1970-2014 (in per cent)

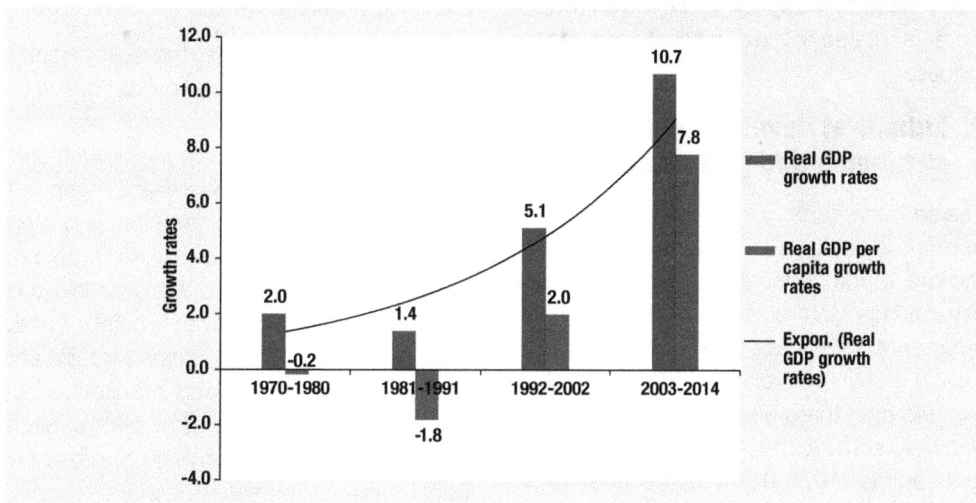

Source: UNCTADstat (accessed on 20 October 2015).
Note: 2014 figures are estimates.

cent in 2004 to 8.8 per cent in 2009. Although recovery has been slow in the period following the financial crisis, annual real GDP growth rates of 12.6 per cent in 2010 and 10.4 per cent in 2013 has turned the country one of the fastest growing economies in Africa. Real per capita GDP rose steadily from $146 in 2004 to $303 in 2014, partly due to a rise in exports earnings of coffee; the latter rose from $223.6 million in 2003/04 to $524.5 million in 2007/08.

1. Overall economic trends: 1970s until the present day

Since the 1970s, the bulk of the country's GDP value added has come from the primary sector which comprises agriculture, hunting, forestry and fishing and accounted for 55.8 per cent of the GDP value added in 1970 (see table 5.1). This rose to 58.4 per cent in 1995 then fell to 45.5 per cent in 2013. Industry's share of GDP value added has declined since the early 1970s; its share stood at 14.4 per cent in 1970 but fell to 11.1 per cent in 1973. Within this, the manufacturing share of GDP value added decreased from 8.9 per cent in 1970 to 5.5 per cent in 1995, and then to 3.9 per cent in 2013.

The share of services value-added has increased over time, particularly in the post-1995 period, rising sharply to 43.5 per cent by 2013. This rise has largely been driven by the growth of the wholesale and retail activities. The rise in health, education, sanitation, recreational, financial intermediation, real estate, public administration and defence services also appeared to have

Table 5.1:   Trends in share of GDP value added by sector in Ethiopia, 1970 to 2013 (in per cent)

Sectors/ Years	1970	1975	1980	1985	1990	1995	2000	2005	2010	2013
Agriculture, hunting, forestry, fishing	55.8	47.5	50.8	43.3	41.1	58.4	47.8	45.2	45.3	45.5
Industry	14.4	16.8	15.5	16.7	16.4	10.4	12.4	13.1	10.4	11.1
Mining & utilities	0.9	1.0	0.8	1.0	1.7	1.9	2.6	2.6	1.9	2.3
Manufacturing	8.9	11.2	10.8	11.2	11.1	5.5	5.8	5.1	4.2	3.9
Construction	4.6	4.6	3.8	4.5	3.6	3.0	4.0	5.4	4.3	4.9
Services	29.8	35.8	33.7	40.0	42.5	31.2	39.8	41.7	44.3	43.5

Source: UNCTADstat (accessed on 24 September 2015).

contributed to the growth in GDP value added by the services sector in the post-1995 period. Many of these investments were facilitated by aid inflows and remittances.

2. Industrial development policy strategies and performance

The earliest efforts to foster industrial development in Ethiopia can be traced back to the late 1950s. Since then, there have been several important shifts that have been instrumental in shaping the performance of local industry. Ethiopia's industrial policy evolved over three distinct periods: the period before 1974, the period during the Dergue regime (1975-91) and the post-1992 period, when the country embarked upon export-led industrialization (Gebreeyesus, 2013). This section presents an overview of these policies and analyzes their impacts on the economy.

a. Industrial development policy focus from late 1950s -1980s

The policy strategies of the first and second development plans (1957-1961 and 1962-1973) were focused on laying the foundation for agricultural-led industrial development through import substitution. A key industrial policy component during the period was the introduction of a decree in 1963 to attract foreign capital into the country.[3] The main focus was on importing intermediate products to produce finished goods in the agricultural and low-technology intensive sectors. However, the rise of production capacities was impeded by the lack of forward and backward intra-industry linkages, and a disconnect between agriculture and other economic sectors (Aregawi, 2005).

From 1974 onwards, there was a marked change in emphasis with more state coordination, which continued until 1991. During this period, Ethiopia's *import-substitution* industrialization strategies were state-led and overseen by a newly created central authority: the National Council for Central Planning. The latter institution engaged in the budget negotiation process with the

Ministry of Finance, in contrast to an earlier approach of minimal intervention by the Planning Commission.

By the end of the 1980s, the policy goal of curbing imports of finished products and intermediate capital goods to help local firms to access markets was implicitly defeated because local firms were mismanaged, and had low-capacity utilization and were uncompetitive. Many of these problems could be traced back to the prior low emphasis on technological upgrading. Faced with the need to import finished goods and manufacturing inputs, and constrained by scarce foreign earnings, Ethiopia's external debt stocks (as a percentage of gross national income) increased rapidly, reaching 25.4 per cent in 1981 and then 69.5 per cent in 1989 (World Bank WDI indicators, 2014).

b. Emphasis on industrial development in the economic reforms of the 1990s

In an effort to address the shortcomings of earlier decades, the overall approach and implementation apparatus of industrial development interventions from the 1990s onwards were dramatically different from earlier interventions between the 1960s and the 1980s. The economic reforms introduced in the 1990s placed a primary focus on private sector-led industrial development and structural reforms along with macroeconomic stabilization and structural transformation. A number of industrial policy interventions were implemented, including measures to privatize public enterprises, the liberalization of foreign exchange (which was pegged at 2.07 units to the US dollar from mid-1970s to the 1980s), trade promotion and enhanced domestic resource mobilization (AfDB, 2002).

c. Industrial development strategies of the 2000s

In the 2000s, the Sustainable Development and Poverty Reduction Programme (SDPRP) of 2002 -2005 focused on market-led industrial development. Under the SDPRP, the industrial policy emphasis was mainly aimed at achieving poverty reduction through an agricultural-led industrialization process, which promoted (IMF, 2004): (i) technical progress in agriculture (by supporting grants to farmers) and agro-processing; (ii) diversification of agricultural outputs; and (iii) greater market interaction.

The SDPRP was followed by the Plan for Accelerated and Sustained Development to End Poverty (PASDEP) in 2006-2010. Under the PASDEP, the government tried to use some of the leverage it had with private sector companies to push forward its agenda for industrial development. It also tried to promote technological and technical progress in agriculture and rural development.

During this time, a more private sector-oriented approach was adopted and a business process re-engineering initiative was launched in 2004 to ensure institutional support to private sector business activities. This approach resulted in the successful streamlining of business process, e.g. the number of days it takes to acquire a trade and investments licence was reduced from 35 days to 34 minutes and 25 days to 2 days, respectively (see Assefa, 2009).

d. The Growth and Transformation Plan and post-2010 policies

Ethiopia's GTP I (2010-2015) aimed to consolidate the gains made under PASDEP, and more particularly to ensure continued stability and sustainable, rapid and equitable growth. GTP I set national targets that should be reached by 2015, these included: 11-15 per cent annual GDP growth; large-scale investments in industrial and agricultural sectors; and the provision of industry-specific incentives, such as duty-free capital goods import for, among others, pharmaceutical and agro-processing industries. To achieve these national targets, several programmes were defined, these included: (i) the development of industrial zones; (ii) capacity building programmes; (iii) university-industry linkages; and (iv) the creation of a centralized R&D and innovation fund. It is expected that each of these programmes will be implemented through

tailor-made projects. They are discussed briefly below.

i. Development of industrial zones

The GTP I envisages the creation of several industrial zones that will catalyze further industrial growth. The first phase of the Bole Lemi Industrial Development Zone has been completed. Preparations are already underway for the second phase of this industrial zone, as well as the development of new sites. For the second phase, $250 million has been sourced from the World Bank to help complete the work (Yaregal Meskir, Deputy Director General, Ethiopian Industrial Development Zones Corporation, 2014).

ii. Capacity building programmes

The capacity building programmes are being implemented in collaboration with the German government and intended to increase the quality of education in Ethiopia. They are also intended to create a technical education stream that is more practice-oriented (i.e. through university polytechnics), as well as strengthen the current system for a higher education at the Masters and PhD levels.

iii. University-industry linkages

In recent years, greater emphasis has been placed on creating and nurturing linkages between universities, research institutes, technical and vocational education and training institutions and industries (see the 2012 STI policy). These linkages were aimed at strengthening earlier initiatives to develop university-industry linkages, such as the understanding reached between the Faculty of Technology, Addis Ababa University and the Ethiopian Manufacturing Industries Association in 2006. This linkage, for example, aimed to nurture closer collaboration between the university and the manufacturing sector to promote technical expertise (Etzkowitz and Roest, 2008).

iv. Creation of a centralized innovation fund for R&D in 2006

In 2006, Ethiopia created a centralized innovation fund for R&D following a revision of the 1993 S&T policy. This innovation fund

Table 5.2:	Trends in number of projects and share of investment capital approved by sector during the GTP implementation (2009/10 to 2011/12)				
Sector	2009/10	2010/11	2011/12	Percentage share of total projects in 2011/12	Percentage share of total investment capital in 2011/12
Manufacturing	1,433	1,294	1,211	21.44	31.12
Agriculture, hunting and forestry	1,342	907	435	7.7	15.92
Real estate, renting and business activities	1,155	1,652	2,694	47.69	15.85
Hotel and restaurants	617	609	271	4.8	8.43
Education	181	143	57	1.01	0.32
Health and social work	99	87	52	0.92	1.92
Construction	942	947	747	13.22	20.38
Wholesale, retail trade and repair services	154	158	22	0.39	0.22
Transport, storage and communication	477	413	101	1.79	0.4
Fishing	8	1	2	0.04	0.02
Mining, and quarrying	9	17	9	0.16	0.11
Electricity, gas, steam and water supply	4	7	2	0.04	4.88

Source: UNCTAD, reproduced from the National Bank of Ethiopia 2011/2012 Annual Report, p. 102.

is financed principally through a tax of 1 per cent of the annual profit of all productive and service sectors.

To implement programmes under the GTP, Ethiopia approved 1,433 projects in the manufacturing sector in 2009/10 (table 5.2). The number of approved projects fell slightly to 1,294 in 2010/11, and dropped further to 1,211 in 2011/12, when it accounted for 21.4 per cent of projects approved for all sectors, second only to projects approved for real estate, renting and business activities.

The GTP proposes a systemic approach to structure institutional support and unlike previous plans, aims to ensure that the public sector becomes more effective and able to work closely with the private sector, civil society organizations and development partners.

3. Overview of Ethiopia's science, technology and innovation policies

Ethiopia's S&T policy framework has evolved in a fragmented manner, and has been somewhat disconnected from national industrial development strategies.

Although national S&T structures were initially created in the 1970s, the commissions, which were charged with the oversight of these structures had to navigate without the benefit of policy frameworks to guide the interactions between various segments of the S&T system. As a result, efforts to integrate S&T objectives with industrial development plans met with little success, and led to difficulties in adequately promoting and coordinating effective support for the enterprise sector.

a. Science and technology policy focus, prior to 2008

Efforts to build a national S&T infrastructure and capability in Ethiopia began as far back as 1975 with the establishment of the Ethiopian National Science and Technology Commission. The Commission set up research councils to oversee, among others, the development of industry and technology, natural sciences, natural resources, education and manpower, S&T, and to create awareness about the importance of S&T. However, as national reviews and interviews conducted for this chapter reveal, the agencies gradually were ineffective in carrying out their respective mandates, e.g. to carry out research, as a result of which effectiveness of policy apparatus and modes of inter-agency coordination weakened over time.

In order to address this, the first national S&T policy was initiated in 1991 after the swearing-in of a new government (Johann and Nelius, 2007). The new government re-established the Ethiopian Science and Technology Commission as an autonomous public institution.[4] In 1993, this Commission prepared the first national S&T policy, which has since then been revised twice, first in 2006 and again in 2010 (UNESCO 2009). This new policy focused more narrowly on capacity building in S&T.

The 2006 revision of the S&T policy once again sought to commit at least 1.5 per cent of GDP on an annual basis to build capacity and deliver 1 per cent of the annual profit of all productive and service sectors to a centralized innovation fund for R&D activities. However, little was achieved under this policy framework in terms of acquiring technologies for the private sector or investing 1.5 per cent of GDP on R&D.

b. Changes in innovation policy: 2008 and beyond

Noting the need to coordinate the policy more coherently, the government created a Ministry of Science and Technology (MoST) in 2008.[5] The vision of the MoST *was* to "entrench the science, technology and innovation capacities of Ethiopia for rapid learning, adaptation and utilization of effective foreign technologies by the year 2022/23". Its *mission* was to "create a technology transfer framework that enables the building of national capacities in technological learning, adaptation and utilization through searching, selecting and importing effective foreign technologies in manufacturing and service providing enterprises"[6].

In 2009/10, in coordination with the launch of the country's GTP I, new initiatives were set up to achieve the national development goals. At the same time, the STI policy was revised to align innovation objectives with the vision of the GTP. The new 2010-2025 national STI policy was developed in 2010 and specifically based on the GTP. The emphasis of this new policy is summarized in box 5.2.

Box 5.2: **Policy emphasis of the national STI policy of 2012**

The new STI policy focuses on realizing seven policy objectives by 2025. These are to:

1. Create a general governance framework for coordinated and integrated STI capacity building;

2. Establish a framework for technology accumulation and transfer;

3. Develop adaptive research that is geared towards rapid technology transfer and adaptation;

4. Develop and commercialize traditional knowledge and technologies;

5. Define the national S&T landscape and to strengthen linkages among the different actors in the national innovation system;

6. Ensure integrated implementation of STI activities with other socioeconomic development programmes and plans; and activities;

7. Institute support mechanisms to progressively increase private sector participation in financing innovation.

Recognizing the need to build capability in the industrial sector, the policy seeks to promote imitation and adoption of new technologies in the first ten years of policy implementation to ensure that, by 2025, the country can have high technological intensity activities in its priority sectors.

The policy stipulates a rise in gross domestic spending on R&D (GERD/GDP) from 0.2 per cent in 2010 to 1.0 per cent in 2015, 1.5 per cent in 2020, and 2.0 per cent by 2025. R&D personnel per 10,000 labour force are also expected to rise from 0.46 in 10,000 in 2010 to one by 2015, five by 2020 and ultimately reach 18 per 10,000 by 2025.

Source: UNCTAD.

C. COORDINATING INDUSTRIAL AND INNOVATION POLICIES FOR FIRM-LEVEL SUPPORT: SURVEY RESULTS

The two sectors chosen for the survey, the agro-processing and pharmaceuticals sectors, are singled out as relevant sectors both in the GTP and 2012 STI policy. Using the same questionnaire as in the two other countries covered in this report, the Ethiopian survey sought to capture how policy changes are impacting on institutional infrastructure and firm-level performance.

1. Sources of technological information

a. The agricultural sector

Agriculture accounted for 42 per cent of GDP in 2014, and coffee remains the country's main source of export revenue, generating between 25 and 30 per cent of total export earnings in 2014 (Tefera and Tefera, 2014). The GTP also acknowledges the specificities of the coffee sector, and seeks to diversify agricultural exports into other areas. It stresses that agricultural and rural development is a central priority and foresees the development of agro-processing industry by 2015.

The GTP sets the following targets for growth in capacity utilization of agro-processing: it was expected to grow from 60 per cent in 2009/2010 to 65 and 70 per cent in 2010/2011 and 2011/2012, respectively, and then to 75 and 80 per cent in 2012/2013, respectively, and finally reach 90 per cent in 2014/2015. Likewise, agro-processing export earnings were projected to rise from $35.2 million in 2009/2010 to $82 million and $144 million in 2010/2011 and 2011/2012, respectively, and then to $150 million and $ 197 million in 2012/2013 and 2013/2014, respectively, finally rising to $300 million in 2014/2015.[7]

Agro-processing firms interviewed in the survey rely on many sources for new knowledge, technology and incremental innovations. The survey found several sources for new knowledge and incremental innovation in agro-processing industry including: (i) hiring of managers and skilled employees, as well as suppliers of equipment or components, (ii) partnerships

---

Box 5.3: The coffee industry in Ethiopia: Building local technological capacity

Ethiopia is home to some of the world's finest coffee. In 2013, coffee accounted for 27 per cent of the country's exports earnings in 2013, and in the 2013/2014 season, coffee exports earnings alone amounted to $841 million USD and similar projections are made for 2014/2015.[8] Although coffee's share in export earnings has declined from 65 per cent in 1995, Ethiopia was the largest producer of coffee in Africa in 2011/2012, and the world's third largest producer of Arabica coffee beans.[9]

Given the enormous potential of the industry to overall economic growth and national development, the Government of Ethiopia is committed to boosting coffee sector productivity and earnings through new initiatives and increased application of new technologies.

First, the Ethiopian Fine Coffee Initiative[10] was launched in 2004 with the objective to own and manage specialty coffee varieties that originated and have been cultivated in Ethiopia. Trademarks were obtained for an umbrella brand *Ethiopian Fine Coffee* and specific brands for Harar, Sidamo and Yirgacheffe. Second, the seven-year Ethiopian coffee quality improvement project (2004-2011) sought to improve access to technologies and facilities for quality improvement. The project was jointly implemented by the Ministry of Agriculture and Rural Development and the International Trade Centre, with the support of other stakeholders. In the course of this project eight laboratories were established in coffee producing areas, new machinery was acquired and staff were trained to improve production techniques.

Agricultural cooperatives are considered to have positively impacted the businesses of about 5 million Ethiopians since the 1990s,[11] and particularly on the agro-processing sector. In the coffee sector, for instance, cooperatives have helped members acquire machinery, such as tractors for mechanized farming practices, and helped to develop skills and in the hiring of professional managers.

Source: UNCTAD, compiled from various sources.

and informal sources; (iii) transfer of technology from parent firm; and (iv) universities and public research institutes.

Some new initiatives have been launched to build the competitiveness of the coffee industry as described in box 5.3.

The survey found that Ethiopia's agro-processing sector has a capacity for expansion beyond the production of coffee (see box 5.4); several companies are now active in producing agricultural goods that are important for the economy (such as leather products and cut flowers) and essential for Ethiopia's vision to achieve food security (including teff). The leather products and the cut flower industry have received some level of foreign direct investment in the past. However, given that agro-processing subsectors, including these are dominated by SMEs, there have been some difficulties in diversifying the sector as a whole, despite the overall progress. Surveyed companies expressed the need for more targeted support to address their needs. The survey also found that other agricultural subsectors that could benefit from product-specific extension services and programmes, as is currently available in the case of coffee. Companies producing leather products, for example, could benefit from programmes that help establish cooperative production and stronger forward and backward linkages.

b. The pharmaceutical sector

There are 18 locally or foreign-owned eighteen pharmaceutical companies in Ethiopia.[12] The development of the local pharmaceutical industry was first recognized as a priority in the GTP, which seeks to "enhance the capacity of existing and newly established pharmaceutical industry to substitute imported drugs, pharmaceutical materials and generate foreign currency earning by exporting the pharmaceutical products". Growth targets in the pharmaceutical industry's capacity utilization have been set to reach 100 per cent by 2012/2013, improving from 30 per cent in 2009/2010 to 50 per cent in 2010/2011, and then to 75 per cent in 2011/2012. The GTP also seeks to increase foreign market earnings of pharmaceutical product exports from $1 million in 2009/2010 to $20 million by 2014/2015. These targets have yet to be met.

The pharmaceutical firms that were surveyed were mostly engaged in pharmaceutical production, including distribution and marketing of imported products, production of secondary material for pharmaceuticals, such as hard-shell capsules, and the production of medical devices and diagnostics. Two local firms, Sino-Ethiop and Cadila Pharmaceuticals, successfully obtained PIC/S certification.[13] The survey showed that the main sources of technological information for local companies were: (i) the firms' own efforts; (ii) hiring of managers and skilled employees as well as suppliers of equipment or components; (iii) partnerships and informal sources; (iv) transfer of technology from parent firm; and (v) universities and public research institutes.

2. Key impediments to upgrading production techniques and performance

The survey results on technological upgrading and the factors facilitating and hindering firm-level competitiveness are summarized under the following headings.

Box 5.4:	Hilina Enriched Foods Processing Centre

Established in 1998, Hilina produces a number of innovative food products, including vitamin A, enriched sugar and iodized salt for UN agencies, NGOs and the general public, both at home and abroad. The aim is to help combat the various forms of malnutrition and other micronutrient deficiencies affecting children and other vulnerable groups.

The demand for Hilina's products inside and outside Ethiopia has been the driving force for innovation and technological upgrading. Hilina is one of many food processing companies in the country. Others include Addis Mojo oil factory, Kokob flour & pasta factory, FAFA food share company and NAS foods; all of these companies have the potential to innovate and expand.

Source: UNCTAD, based on field interview with the company.

a. Greater support to develop process and product technologies and innovation capacities is needed

Companies stressed the need to have adequate support to develop technology and innovation capacity to help improve their productive capacity as a whole. In the agro-processing sector, the survey showed the need to innovate and expand (see box 5.4). In the pharmaceutical sector, the survey showed that local manufacturers produce only 90 of the 300 drugs on the national essential drug list. Companies producing disposable syringes and other medical supplies can only supply 20 per cent of the country's needs for these products.

Local companies have severe limitations as a result of low technological capacity:

(i) Almost none of the local companies are able to produce pharmaceutical products that meet WHO prequalification standards, as a result of difficulties in upgrading their products and processes (WHO cGMP).

(ii) Local pharmaceutical firms only produce 5 per cent of all the intermediate products being used, although there are some outstanding examples of such production (see box 5.5). The rest are imported to facilitate local production of pharmaceuticals, which drives up the costs of drugs produced.

(iii) Lack of skills hinders the ability of pharmaceutical companies to engage in reverse engineering and learning activities.

(iv) Most importantly, there is no local capacity to produce active pharmaceutical ingredients (APIs), which are essential in promoting local production capacity in the pharmaceutical sector.

The survey also showed that training new staff is also quite a low priority in the current context, as companies only spend between 0.5 and 5 per cent of the total payroll to meet the training needs of their personnel. The survey also found that there were insufficient numbers of public sector agencies/centres of excellence to conduct applied research of the kind that can feed into the local pharmaceutical sector to help them develop skills, such as those needed in reverse engineering. Firms also pointed out that there was the huge human resource gap, which made it difficult to meet industry development targets, and a need to have more university graduates with relevant training coming into the industry.

On the whole, the survey and the field interviews show that lack of access to and sharing of R&D facilities continue to hinder the ability of local firms to take advantage of opportunities both within Ethiopia and in other in emerging markets. A total of 33 per cent of the firms in both sectors indicated that they had benefited greatly from access to and sharing of R&D facilities.

b. The need to exploit emerging markets and build competitive industry strategies

The survey showed that firms continued to focus on domestic market opportunities; however, a few companies, particularly cof-

---

Box 5.5:   Sino-Ethiop Associate (Africa) Private Limited Company

Sino-Ethiop Associate (Africa) Private Limited Company is an Ethiopia-Chinese joint venture company established in March 2001. Ethiopia has a 30 per cent stake in the joint venture. The company manufactures and markets hard gelatin capsules for the local market and exports to other African countries, such as the Democratic Republic of Congo, Kenya, South Africa, Sudan, Zambia and Zimbabwe. It also produces medical devices and packaging materials. Of its total production, company executives estimate that about 70 per cent of production is sold in the local market, while 30 per cent is exported.

Sino-Ethiop has an annual production capacity of 1.2 billion capsules. Despite this capacity, the company is not able to meet the increasing demand for their products and is planning to expand.

Source: UNCTAD, based on field interviews with the company.

fee producers, have ventured into markets beyond Ethiopia. Among exporting firms, most acknowledged that export demand has helped to shape their innovation strategies. However, agro-processing companies that did not have a particular market niche for their products overseas found it difficult to export their products or consider new technological breakthroughs in their production, including producers of Teff, a widely grown crop in Ethiopia. Industrial policy instruments may be appropriate to enable agro-processing companies to produce and export their products, in part to prevent extreme dependence on coffee exports.

c. The need to create forward and backward linkages

To a certain extent, surveyed firms acknowledged that the largest impediment was perhaps the lack of forward linkages (with other sectors and production chains that could help the upgrading processes) and backward linkages (with the agricultural outputs). Inefficiencies in backward linkages often resulted in delays and stalled productions, with associated costs for the agro-processed goods; an effective solution to this problem, as frequently suggested in the interviews, was to establish and empower agricultural cooperatives.

d. The need to closely align industrial and innovation policies

The survey pointed to several policy gaps that needed to be urgently addressed in three specific areas:

(i)   Improved government's support and effort to promote technology transfer and an enabling R&D environment;

(ii)   Business-friendly Customs procedures and export-import policies; and

(iii) Easier access to land for business purposes.

The survey also identified policy areas where the government or other institutional support structures could be critical in formulating new strategies. Between 53 to 75 per cent of enterprises interviewed pointed out that it is important for government agencies to promote technology transfer and create a more enabling R&D environment and access to land (ease registration cost and procedures), and that this was extremely critical to promote innovative capacity.

e. Improving exports-import procedures and the general business environment

The survey also showed that more can be done to make current Customs procedures more business-friendly. Analysis of survey findings on policy constraints that impact industry efforts to build technology capabilities showed that a majority (53 per cent) of those surveyed identified Customs procedures and export-import policies as having a severe negative impact on their business operations. Also, the survey findings showed that restrictions on licensing arrangements, local duties, access to land and municipal regulations do not impact day-to-day business operations in a serious manner. Respondents emphasized that curbing official corruption and other regulations could promote innovation and industrial development.

f. Promote access to finance

Lack of financial support was identified as one of the main constraints on the ability of enterprises to take advantage of opportunities and innovate. Given the risky nature of innovative activities, local firms reported that obtaining domestic bank loans for innovative activities are not easy, and often lead to interesting ventures being abandoned.

D. CONCLUDING REMARKS

Despite the policy traction in the realm of industrial and innovation policies, Ethiopia still faces some constraints, which prevent its industries from realizing its potential. The analysis in this chapter has shown that many of these shortcomings have serious repercussions for achieving the goals set out in the GTP. Although progress has been achieved in both the pharmaceutical and agro-processing sectors, the targets set out in the GTP have not been met as the analysis in the previous section has shown.

Building policy coherence and coordination is a slow process, but at the same time, it is important to note the key areas where further action is required. The chapter points to the following important constraints that need to be addressed in the next GTP in order to consolidate and extend the momentum achieved up until now. These are:

(i) Coordination between industrial and innovation policies:

Perhaps the most important finding of the sectoral surveys was that the companies would be best equipped to expand, technologically upgrade and compete if there was more policy coherence and coordination at the implementation level between the industrial policy vision and GTP targets and the national STI policy. The survey results also lend strength to the conclusion that more immediate action is required to equip agencies, such as the Food and Beverages and Pharmaceuticals Industry Development Institute (FB-PIDI). These agencies are crucial as they can help steer firm-level performance in the right direction by providing industry-relevant services (see box 5.6)

(ii) Physical, trade and transport-related infrastructure costs:

Trade and transport-related infrastructure are crucial for innovation and industrial competitiveness. Ethiopia has seen an improvement in this area since 2007 but there are still large gaps. Access to electricity increased slightly from 23 per cent of the population to 23.3 per cent between 2010 and 2012. Mobile cellular subscriptions (per million people) rose from 5,391 in 2005 to 315,939 in 2014. Fixed broadband internet subscribers (per million people) also rose from 0.8 in 2005 to 4,883 in 2014. However, such infrastructure is still not readily and easily accessible to firms at a low cost.

(iii) Knowledge infrastructure and skills development:

Ethiopia is currently implementing its Education Sector Development Programme IV (2009/10 to 2014/15). This programme emphasizes the development of technical and vocational education and training, as well as the overall knowledge and human resource infrastructural development in the country. As of 2014, there were 31 public universities, 59 non-government higher educational institutions and 29 colleges of teacher education in the country. Public R&D spending is also rising, although not to the extent envisioned in the GTP. According to the MOST, 0.6 per cent of GDP was invested into R&D in 2014, and the number of researchers in R&D (per million people) rose from 21 in 2005 to 42 in 2010 along with an increase in the number of scientific and technical journal articles from 88 in 2005 to 170 in 2011. Despite this, there is a need to ensure that the increased R&D spending is prioritized towards industry-relevant research.

(iv) Collaboration and network for access to new technologies and innovations for industrial development:

The National STI policy of Ethiopia is a relatively new document; however, this policy needs to be considered in the context of the country's experience over the past three decades. Comparing the current document with its predecessors, what stands out is that several of the targets set out in the current STI policy are similar to those in the 1993 S&T policy. Among the targets being implemented are a 1.5 per cent investment of GDP into R&D, promotion of technology transfer, strengthening of public research institutes, and greater collaborative networks. A helpful next step would be to conduct a policy review to assess previously encountered difficulties, and to take steps to avoid the same problems in the future. Furthermore, there is an urgent need to have new institutions that interface research with product development. A recent effort to bridge this gap is the Food and Beverages and Pharmaceuticals Industry Development Institute (see box 5.6).

Box 5.6:	Food and Beverages and Pharmaceuticals Industry Development Institute

The Ministry of Industry in Ethiopia set up the Food and Beverages and Pharmaceuticals Industry Development Institute (FB-PIDI) to act as a one-stop shop to assist the food processing and the pharmaceuticals sectors. The effective functioning of this institution is critical to boosting the capacity of these two sectors, particularly with respect to upgrading production facilities, and to provide advice and support on boosting capacity; FB-PIDI is also expected to provide research, laboratory and testing services to the industry. This is a first initiative of its kind in Ethiopia, and its future success and capacity will largely depend on the infrastructure and manpower it will benefit from. At the time of the field investigation, the agency was limited by funding and weaknesses in human skills and infrastructure that policymakers have acknowledged and which are expected to change soon.

Source: UNCTAD, based on field interview with the company.

(v) Channeling knowledge flows, internal and external:

Business does not operate in a vacuum: Company decisions are largely influenced by opportunities within domestic markets. This has to be complemented with greater knowledge flows accruing from collaborations with national and foreign firms and governmental agencies in order to set up a broader base for innovation and industrial competeveness. Data show that the share of enterprises with an internationally recognized quality certification (e.g. PIC, GMP or ISO certification) increased from 4.2 per cent in 2006 to 13.6 per cent in 2011. Likewise, the share of enterprises using technologies licensed from foreign companies also rose up from 4.2 per cent in 2006 to 42.7 per cent in 2011. However, collaborative linkages focusing on local learning and local innovation content still need to be fostered further.

(vi) Business environment and innovation cost:

According to the World Bank's study on Doing Business (World Bank, 2014), the total tax rate as a percentage of profit is more favourable to enterprises in Ethiopia where the rate is 31.8 per cent compared to the sub-Saharan African average of 46.2 per cent. On the whole, Ethiopia has a better 'ease of doing business' and innovation environment for industrial development than many other African countries. The improvements in this regard should be coordinated with R&D and skills development to foster technological investment at the firm-level.

(vii) Delays in delivery times and high import-export cost:

Industry also pointed to delays in delivery times hindering their ability to export or import. While it takes less than one month to export or import to East African countries, such as Kenya and the United Republic of Tanzania, it takes on the average one and half months to do the same in Ethiopia. The cost to import per container of inputs in 2012 was around $2,660 in Ethiopia compared to for example, $1,200 in India. Exporters and importers need better conditions in order to maximize gains from trade.

(viii) Promote overall export orientation:

The local industry is still inward looking and its focus on learning opportunities, both internal and external, need to be promoted. In this regard, the STI policy framework needs to actively engage in promoting collaborative networks in the country and outside.

(ix) Embed the implementation of the STI policy in the GTP:

While the country has achieved laudable successes by integrating the GTP and the new STI policy, the implementation of STI policy has not yet been embedded within the broader GTP framework. This is highlighted in the difficulties faced by local firms to invest in technological learning and navigating through the industry development and promotion landscape. Acquisition of new knowledge is necessary but that may not be sufficient for sustainable industrial development if not complemented by innovation and creativity within industries.

(x)   Take local industry characteristics into account:

Similar to the situation in the United Republic of Tanzania, the STI policy and the GTP need to be revisited to better incorporate local industry characteristics and ensure that the day-to-day activities of local firms are better integrated. Currently, although recognized in the STI policy, there is a need for clear institutional mechanisms to implement the various policy objectives, and to promote technology transfer, access to finance, joint ventures for production, value-addition in agro-processing beyond coffee, mechanisms for policy feedback and sector-based associations with enhanced links to government bodies. Going forward, there will be the need to interlink and apply innovation focused industrial policies and those of the industry focused innovation policies to create synergies that could help address the industrial development needs of Ethiopia.

The second phase of the GTP should build on the current momentum to introduce innovation focused industrial policies with a focus on promoting strategic partnerships and informal sources of technologies along with intra-industry (knowledge flow within industry) and inter-industry (knowledge flow between industries) linkages. The role of universities and public research institutes will be critical in this process. In addition, technology licensing, technologies from joint venture R&D arrangements and technologies for reverse engineering could be crucial.

NOTES

1. Ethiopia STI policy draft, see: https://www.healthresearchweb.org/files/Ethiopia_National_S,T&I_ Policy_Draft.2006.pdf

2. Data for Eritrea are not reflected as of 1992 as it gained independence.

3. Ethiopia Investment Code (1963).

4. Proclamation No. 91/1994 on Ethiopian Science and Technology Commission Establishment. http://www.wipo.int/wipolex/en/text.jsp?file_id=175299

5. Proclamation No. 603/2008.

6. Ethiopia Ministry of Science and Technology website: http://www.most.gov.et/index.php/ministry/ about-the-ministry/mission-vision

7. Federal Republic of Ethiopia Growth and Transformation Plan 2010/11 - 2014/15. Volume II: Policy Matrix. Ministry of Finance and Economic Development, 2010, Addis Ababa.

8. Ethiopian Coffee Exports to Hit Record in 2015. http://ethioagp.org/ethiopian-coffee-exports-to-hit-record-in-2015/

9. World's Top 10 Coffee-Producing Countries in 2010-2011. Bloomberg News. http://www.bloomberg.com/news/2011-08-19/world-s-top-10-coffee-producing-countries-in-2010-2011-table-.html

10. The Coffee War: Ethiopia and the Starbucks Story. http://www.wipo.int/ipadvantage/en/details.jsp?id=2621

11. Great Programs in History: Agricultural Cooperatives in Ethiopia. http://www.acdivoca.org/site/ID/FeatureGreatProjectsinHistoryAgriculturalCooperativesinEthiopia

12. These companies include: Addis Pharmaceutical Factory; Epharm; Cadila Pharmaceuticals; Rx Africa; Fawes Pharmaceuticals;East Africa Pharmaceuticals; National Veterinary; Pharmacure; Sino Ethiop; Asmi Industry; Fanus Med Tech; MOAB; ARFAB Engineering Medical Equipment Manufacturing; Access Bio.; Medsol Pharmaceuticals; Tulips Cosmetics & Pharmaceutical Manufacturing; Julphar; and Samed.

13. The term PIC/S refers to the Pharmaceutical Inspection Convention and Pharmaceutical Inspection Cooperation Scheme. These two international instruments enable cooperation on matters of promoting good manufacturing practices between countries and pharmaceutical inspection authorities.

PARTNERING FOR DEVELOPMENT: HARNESSING THE SYNERGIES BETWEEN INNOVATION POLICIES AND INDUSTRIAL POLICIES

6

CHAPTER VI
PARTNERING FOR DEVELOPMENT: HARNESSING THE SYNERGIES BETWEEN INNOVATION POLICIES AND INDUSTRIAL POLICIES

A. INTRODUCTION

African countries have reached a defining point where stocktaking is not only necessary but also vital, particularly as they prepare to address the new development goals contained in 2030 Agenda for Sustainable Development. It is widely acknowledged that sustainable development rests more broadly on stable industrial development of a kind that can deliver better livelihoods and eradicate poverty, as reflected in several goals of the 2030 Agenda for Sustainable Development. Goal 9 of the SDGs encapsulates the dual objectives of promoting inclusive and sustainable industrialization and fostering innovation. In this context, innovation and industrial development are highly relevant from an African perspective, and are being extensively supported by newer literature emerging on the topic. Important results include:

(i) Innovation policies are relatively new and often not well implemented in a large number of countries.

(ii) Innovation systems suffer from many shortcomings, many of which continue to affect their effectiveness.

(iii) In the past, the industrial development experiences of African countries have been shaped by similar concerns and strategies in the African region (e.g. import substitution strategies, trade liberalization, shift to export promotion, and less emphasis has been placed on promoting technological learning, etc.) (Noman et al, 2012).

(iv) In the instances that industrial development strategies have focused on technological change, there have been several problems related to proper coordination (Noman et al, 2012; Cimoli, Dosi and Stiglitz, 2009).

(v) Most sub-Saharan African firms are family-owned and small-sized; this hinders their financial resources and capacity to acquire modern technologies. Moreover, scarcity of human resources, brain drain, weak governance in technology transfer, and a weak enabling environment all serve as other main barriers for innovation (UNECA, 2014).

(vi) Fostering firms through a supportive policy environment will be crucial to promoting sustainable industrial development in the region.

Starting with a detailed review of the industrial development strategies and policies, and the STI policies of a large number of African countries, this report has proposed an analytical framework to consider the overlapping domains in the two policy frameworks, along with a set of principles that could help structure the interactions in a complementary manner and help steer industrial development through an innovation-oriented industrial policy. Chapters III-V of this report analyzed the linkages between industrial and innovation policy frameworks within countries by collecting primary data collected through a purposive survey designed to understand policy gaps (in policy history, conceptualization and implementation) and their impact on firm-level performance.

An analysis of national experiences with industrialization and the varied institutional backgrounds of the three countries (Nigeria, the United Republic of Tanzania and Ethiopia) shows that developing and implementing policy processes that shape the interaction of industrial development and innovation policies right are not to be under-estimated, particularly due to the significant impact they have on firm-level performance. The historical consideration of the industrialization experiences allows us to draw lessons from the past and apply them to the present and the future.

The findings of these chapters highlight the continuing challenges of fostering technology-led industrial development in countries. At a fundamental level, the analysis illustrates how experiences of industrialization have had limited success in the countries as a result of the lack of integration of technological learning and STI issues.

This final chapter presents key results from a more general perspective, and in its final part proceeds to evaluate the most critical factors of the industrial policy-innovation policy interface and to provide policy recommendations.

B. GENERAL FINDINGS

There are at least three general findings that are of relevance from the foregoing chapters of the report, and also help discuss the results from a broader perspective. Almost all countries in the African region, including the three countries studied in this report, are currently at a policy and developmental stage where industrial development through technological change is a central, if not the most important, priority (See tables 2.1 and 2.2 of chapter II). Not only is there a policy transition towards that end, i.e. the field surveys reflect the political commitment that exists with regard to enacting elaborate industrial policy frameworks and revising S&T policies and re-orient them towards policies dedicated to innovation, in terms of STI policies. Thirdly, the private sector in the African region (particularly in sub-Saharan Africa) is in dire need of greater support.

1. Countries have elaborate industrial policy frameworks

All three countries have significant experience in enacting industrial policy frameworks. Starting out with import substitution strategies in the 1960s and 1970s, the three countries have progressively transitioned to different kinds of export-oriented strategies from the late 1980s onwards.

In the case of Nigeria, the National Industrial Policy of 1998, is an elaborate policy that was aimed at resuscitating the industrial sector through structural diversification, promotion of new sectoral activities, and increasing the manufacturing value-added of products and export promotion. Along with that, in 2002, Nigeria also launched its Vision 2010 aimed at making Nigeria Africa's leading economy by 2010. Later in 2010, Nigeria adopted its Vision 20:2020, which is currently being implemented through a series of medium-term plans: the first of which was developed for the period of 2010-2013, and the second and third for the periods of 2014-2017 and 2018-2020, respectively. Likewise, Nigeria's multifaceted 'Industrial Revolution Plan' launched in January 2014 seeks to provide a national roadmap for industrialization.

In the United Republic of Tanzania, the multipurpose industrial development policy (SIDP, 1996-2020) was formulated to promote "indigenous entrepreneurial base through orienting the education policy and strategy to emphasize technical education, including strengthening of vocational training institutions and entrepreneurship development" (p. 13-14). This was followed with the Tanzania Development Vision (TDV) 2025, which was launched in 1999 to transform the country into a semi-industrialized economy and attain middle-income country status by 2025. To this end, the country adopted the first National Strategy for Growth and Reduction of Poverty (NSGRP I, 2005-2010), and the second National Strategy for Growth and Reduction of Poverty (NSGRP II, 2010-2015) to help achieve TDV 2025. Along with that, the Integrated Industrial Development Strategy

2025 (IIDS) was enacted in 2011 to guide the process of resource-based industrialization with emphasis on industrial cluster development, PPPs and SEZs.

Ethiopia's overarching industrial development strategy is Ethiopia's Growth and Transformation Plan (GTP I, 2010-2015), which is aimed at achieving national targets, such as 11-15 per cent annual GDP growth and large-scale investments in the industrial and agricultural sectors. The GTP I seeks to emphasize programmes, such as development of industrial zones, industry capacity building and university-industry linkages to help achieve these national targets. With GTP I expiring in 2015, plans are in hand to enact a second Growth and Transformation Plan (GTP II), which build on the industrial achievements and successes of GTP I.

2. Countries have elaborate STI policy frameworks

In line with broader trends observed in the developing world, as can be seen in chapters I and II of this report, all three countries have transitioned from S&T policies to STI policy frameworks.

Nigeria's 2011 STI policy is broad and marks a substantial shift from its earlier emphasis on S&T policies. The policy mission is fairly extensive with a focus on building a nation that harnesses, develops and utilizes STI to build a large, strong, diversified, sustainable and competitive economy. The general objective is to "build a strong science, technology and innovation capability and capacity needed to evolve a modern economy" (page 1). These broader objectives are linked to specific industry targets, such as the production of solar cells, ICT industry development and applications, as well as the development and application of nanotechnology, chemical technology and biotechnology. All these sectoral priorities are expected to feed into the existing industrial hub (Sheda Science and Technology Complex/ SHESTCO).

Nigeria also emphasizes strategies to popularize and inculcate STI culture and to mainstream STI in development programmes focused on women. One important development is that the new policy is linked to job creation, innovation and national development at large.

In the case of Ethiopia, the STI policy of 2012 is the guiding framework, which envisions creating a national technology transfer and innovation framework to help build capabilities that would enable rapid learning, adaptation and utilization of effective foreign technologies by the year 2022/23. The policy has seven important objectives including the plan to "establish and implement a coordinated and integrated general, governance framework for building STI capacity" (STI Policy, p. 4).

The United Republic of Tanzania is currently in the process of enacting a new STI policy framework, which envisions the promotion of innovation and technology development, transfer and commercialization.

3. Enterprise support policies are the weak link

A third, significant result is that local firms across the three countries are operating in a highly constrained institutional environment, both in terms of industry and innovation support. The field surveys shed light on some aspects of the institutional environment and what the deficiencies are.

Firstly, a large number of the local firms are MSMEs that often operate in the informal sector and on the fringes of the local economy. The only exceptions to these are a handful of large-scale firms, which are able to acquire sophisticated technological machinery for manufacturing purposes. Despite this, micro- and medium-sized firms make large contributions to the economies of the three countries surveyed. In Nigeria, about 17.3 million MSMEs are estimated to have contributed about 46.5 per cent of the country's GDP in 2010. In the United Republic of Tanzania over a million MSMEs are estimated to account for 95 per cent of businesses; these firms generate 30-35 per cent of GDP and are responsible for 40 per cent of total employment in 2009.[1] Similarly, SMEs have accounted for 3.4 per cent of

GDP and over 52 per cent of the industrial sector contribution in Ethiopia since 1993.[2] Survey results confirm that for a lot of these firms, small changes in the institutional support structure would go a long way to improve productivity and remain afloat.

Technological upgrading capacity and financing remain two key obstacles to day-to-day company operations. The survey results show that firms operating in the medium- and high-technology sectors are often engaged in activities that are far removed from production or process improvements, product design or reverse engineering activities, i.e. not engaged in activities that can be classified as learning or technological upgrading. For example, many ICT companies in Nigeria and the United Republic of Tanzania were providers of ICT services, including Internet providers and call centres.

Most importantly, survey results show that STI policies have a rather limited effect on raising productivity and increasing the upgrading and expansion of smaller firms and the informal sector (see also Benjamin and Mbaye, 2012). In the United Republic of Tanzania, for example, the Tanzanian National Business Council has acted as a forum for dialogue between the private and public sectors, but this has generated mixed results. There is strong political will to use this council for policy implementation (Page, 2014), but insufficient coordination in the current context of policy framing and implementation has resulted in a weak focus on local industrial development.

C. INDUSTRIAL POLICY-INNOVATION POLICY INTERFACE: WHAT MATTERS

Moving from the general findings to more specific aspects, countries continue to face several common constraints, many of them due to historical path dependencies following the manner in which institutional frameworks have evolved, as well as the alignment of industrial policies with innovation policies. These limitations correspond to the issues highlighted by chapter II of this report that elaborated on a set of five principles in the industrial policy-innovation policy interface (box 6.1).

1. Gaps in policymaking structures exist

In all three countries, as is the case with a large number of other African countries that are also reviewed in the report (see chapter II), national STI policies either evolved much later (at least two decades after the industrial development policies were enacted), or evolved in parallel with little or no coordination with established industrial development frameworks.

The report finds that within countries, a predominant issue is *where industrial policy is placed*, and how it is articulated. In the case of a large number of developing countries, policies for industrial development are not usually articulated as industrial policies, but rather as industrial development strategies, or as national visions, or as part of recurring national developmental plans aimed at

---

Box 6.1: **Five guiding principles in aligning the industrial policy-innovation policy interface: A recap**

1. Identify and eliminate policy redundancies in the policy conceptualization and policy making structure.

2. Promote policy coherence and policy competence.

3. Use resources carefully.

4. Develop capacity for proper policy evaluation and monitoring.

5. Coordinate the policymaking processes closely vis-à-vis their impact on the business and enterprise environment, and promote the engagement of the private sector.

Source: UNCTAD.

facilitating overall development and economic transition.

If countries enact national visions that include industrial policy objectives (which is the case not only in Ethiopia, Nigeria and the United Republic of Tanzania, but also true for a large number of other African countries), it needs to be borne in mind that such national vision statements generally have a broader scope than just promoting industry, and often tackle issues of poverty, youth, environment, employment and urbanization. In several countries, industrial development objectives are embedded in their national development plans, and are often recurrent on a term-by-term basis.

Therefore, although such visions or strategies encapsulate the main industrial objectives or goals, there is a need to have clear roadmaps to achieve these visions, with accompanying targets, so that these can be linked to a policy implementation mechanism on the one hand, and to STI and other policies (covering areas such as trade, investment, and development) on the other.

Another reason for the gaps in policymaking is that a large number of industrial development strategies are one-dimensional: they target overall industrial development and an increase in per capita GDP growth rates, or a rise of specific sectors. The focus should instead be on closing the productivity gap, i.e. how to ensure greater returns from productive activities. This leads to gaps in policymaking, including a neglect of:

(i) Technological and technical support systems required for the growth of sectors;

(ii) Links between the human skills requirements of the various sectors with enhanced performance projections;

(iii) A clear articulation of how the higher GDP spending on R&D will form part of public sector assistance to technological upgrading, e.g. the establishment of common industry services, technological incubation, industrial research labs, etc.

2. Policies suffer from inconsistencies and often, overall incoherence

A key issue that stands out is that sophisticated policies are not sufficient. While industrial development strategies in the selected countries recognize the importance of technology-led growth, and whereas all STI frameworks recognize the importance of coordinating with industrial policy, the same historical patterns of lack of coordination between innovation and industrial policy frameworks persist. Countries have tried to tackle these issues by providing for common goals or missions in the two policy frameworks, but policy incoherence not only occurs at the stage of policy articulation, and is also often deeply rooted in policy implementation processes.

The country chapters help to illustrate the main finding of the analytical framework, namely that it is crucial that *policy processes* are clearly laid out. Specifically, the findings show that even elaborate policy frameworks on STI policy and industrial development need to be accompanied by policy consistency and coherence at the levels of:

(a) Policy conceptualization and design;

(b) Policy implementation and coordination

a. *Policy incoherence in the conceptualization of the two policy frameworks*

A number of reasons explain the existence of policy incoherence and inconsistencies. The country chapters show that they could be the result of ineffective policy transitions (where countries embark on changes in policy, but remain incomplete and lose momentum as a result of changing political leadership at different levels of governance), institutional inertia and resistance, or a lack of policy competence to foresee and avoid overlaps.

i. Incoherence as a result of ineffective or slow policy transitions

Moving towards an innovation policy is a challenging coordination task and involves more than just providing a regulatory framework. In reality, although a wide variety of policies emphasize 'innovation', field investigations found that while some policies seek to fundamentally chart new ground, in other instances the policies often make reference to 'innovation', but are not comprehensive enough to tackle the challenges of fostering innovation. Furthermore, additional difficulties arise when policy processes are not followed through, or maintained through changes in the political environments in countries.

The same difficulty holds true for policies in industrial development. For instance, in the case of the United Republic of Tanzania, the Sustainable Industrial Development Policy (SIDP) 1996-2020 reserved the right of the government to invest in critical sectors. As a part of this, the Tanzania Mini Tiger Plan 2020 was launched in 2005. But as of 2005, this was replaced by the National Strategy for Growth and Poverty Reduction 2005-2010, and in 2011 followed by a new Integrated Industrial Development Strategy 2025. It remains unclear what the link/continuity between the two industrialization strategies are, particularly given that the philosophies of the two strategies are very different. As opposed to the 1996 Industrial Development Policy, which sought to replicate the successes of the East Asian economies by investing in particular sectors, the new 2011 strategy advocates resource-based industrialization. These kinds of sudden shifts that do not help foster industrialization as a continuous process lead to policy inconsistency and incoherence simply because they do not offer a consistent and reliable support to the process of industrial transformation.

ii. Incoherence due to institutional resistance and inertia

The field interviews and surveys in all three countries shed light on the fact that policy and institutional history matters. Historical analyses of the evolution of policies and implementation mechanisms conducted in the chapters shows that agencies implementing these mandates operate within weak, unaccountable implementation processes. Such inter-agency rivalries exac-

erbate policy coordination issues and have led to a large-scale neglect of the private sector. In almost all countries surveyed, private sector enterprises considered that existing policy frameworks and the actions of implementing agencies operated at a distance from them, making little attempt to liaise and understand the constraints they faced or tried to alleviate them. The field investigations showed that entrenched institutional habits and practices were hard to change, and explained why newer more collaborative modes of interaction were not emerging, despite policy mandates. Policies on industrial development, if they are to be coherent with innovation policies, should seek to address the operative mandates of agencies to promote a change in mindset.

iii. Incoherence due to insufficient policy competence/ policy oversight

Another set of coordination issues relate to targets included in both industrial development policies and innovation policies. These are often designed, and intended to be impacted upon by the policies differently. For example, in Ethiopia, the STI policy aims to "develop, promote and commercialize useful indigenous knowledge and technologies". To promote this, an assessment would first be needed on whether the *sui generis* system created by the Ethiopian 2006 Proclamation on Access to Genetic Resources and Community Knowledge and Community Rights could help protect useful indigenous knowledge and technologies. In other words, IPR protection has to be integral part of the indigenous knowledge commercialization process. But what seem to be missing in the objectives are strategies to create STI policy awareness at all levels of government including the Cabinet and Parliament, as well as strategies to build innovation culture among businesses, the youth and society at large. Similarly, one of the projects under the GTP is the setting up of industrial parks, but these are expected to perform the function of acting as hubs of foreign direct investment and to leverage technology transfer of the kind outlined in the country's STI policy. This once again calls for a strategic approach to the coordination of policy implementation between the ministries and agencies involved in implementing the mandates on industrial development, investment and STI. However, the lack of policy competence and incentives among agency employees often leads to very minimalistic interpretations of these mandates.

Assessing the successes and difficulties faced by the countries in this report, it is advisable to implement the following recommendations in order to avoid policy incoherence that can arise in conceptualization and design:

(i) *Policy vision, mission and objectives should be closely aligned.* A review of all the initiatives, as can be found in tables 2.1 and 2.2 of chapter II and the country chapters, lend strength to the conclusion that a close alignment of industrial development and innovation policies is often an elusive goal in countries. Oftentimes, even the missions, or the STI objectives covered in industrial policy are not the same as the objectives of the STI policy itself (see previous point), leading to policy incoherence and confusion. A case in point is the *vision* and *mission* segments of Ethiopia's current innovation policy, which appear to focus on building the capabilities to exploit foreign technologies but do not seem to emphasize the establishment of industry-friendly innovation system, as envisaged in the policy. It is imperative for the policy vision and mission to be aligned with the policy *objectives*, which seem to focus on building broader innovation capabilities, including the development and exploitation of local indigenous knowledge and technology.

(iii) *Emphasis should be placed on developing local linkages and unlocking learning potential*: Although STI policies clearly lay down the broader vision to build capacity, fostering an innovation ecosystem calls for emphasis on the creation of innovation and entrepreneurship culture with express links to industrial development. It is therefore necessary to promote entrepreneurial programmes, align academic curriculum with entrepreneurial needs, and introduce entrepreneurship classes at schools and institutions of higher learning for the effective application of new technologies and innovation for industrial development.

(iv) *While enacting new policies, there is a need to clearly link them with existing initiatives and agency mandates:* As seen in the country chapters, policymakers are aware of the need to review existing policies and agency mandates, but change is usually slow, leading to policy ineffectiveness.

(iii) There are two critical reasons why this should occur in tandem with the policymaking/revision process is critical for at least for two reasons: Firstly, because previous policies often have agency mandates that call for a review as a new policy is introduced; and secondly, to ensure that the institutional framework embodies the changes in a dynamic and efficient way. An example in this regard is Nigeria's STI policy, which has rightly underlined the need for the existing institutional and legal framework to be restructured in such a way as to strengthen national innovation capacity as an essential first step for a strong national innovation system. However, the difficulty is that since its inception in 2011, the policy has been implemented within an institutional setting that was created for a different purpose. Furthermore, several older policy directives that were set out for review have yet to be considered. For example, there is an indication in the new STI policy that the National Science and Technology Act, CAP 276 of 1977 and the Federal Ministry of Science and Technology Act No 1, 1980 would be reviewed. The mandate of the National Office for Technology Acquisition and Promotion, which was created in 1979, also needs to be reviewed. Reviewing these policies and others, including the 1986 Federal Universities of Technology Act and 1987 National Science and Technology Fund Act, will be critical to ensure that these two Acts are in line with the objectives of the current STI policy.

- Secondly, reviewing policy mandates is very important in ensuring that national resources (financial and human skills related) are used efficiently.

b. Policy incoherence in the implementation process

A second form of policy incoherence is when the frameworks are overarching but not accompanied by a concrete implementation plan. However, in many other cases, policy frameworks are accompanied by implementation mechanisms, but several shortcomings have prevented them (to a different extent in the three countries) from achieving an impact. A key issue (already raised in the previous point) is that in the absence of stocktaking and attempts to streamline the institutional apparatus, many public sector agencies have mandates to implement the policies and related incentives, and this can lead to duplication. When the policy framework is not completely consistent or accompanied by clear implementation mechanisms, the country analyses show that there is no clarity at the policy implementation stage as to which of the existing agencies should implement the mandates contained in the policy framework and how they should be implemented.

The situation is much more drastic than one could imagine. Box 6.2 contains the national agencies with mandates to implement the incentives within industrial development policies and STI policies in each of the three countries examined in this report.

Box 6.2:	National agencies/parastatals with mandates to implement industrial and innovation policies

1. National agencies/ parastatals in Nigeria:

In Nigeria, the Federal Ministry of Industry, Trade and Investment (FMTI) was created to "formulate and implement policies and programmes to attract investment, boost industrialization, increase trade and exports and develop enterprises". FMTI has 14 parastatals, namely the Abuja Securities & Commodities Exchange; Bank of Industry (BOI); Consumer Protection Council; Corporate Affairs Commission (CAC); Industrial Training Fund (ITF); Financial Reporting Council of Nigeria; Nigeria Export Processing Zone Authority (NEPZA); Nigeria Export Promotion Council (NEPC); National Automotive Council (NAC); National Sugar Development Council (NSDC); Oil & Gas Free Zone Authority; Standard Organisation of Nigeria (SON); Small & Medium Enterprises Development Agency of Nigeria (SMEDAN); and the Nigeria Investment Promotion Commission (NIPC).

The Federal Ministry of Science and Technology in Nigeria has the following 17 parastatals, namely: Energy Commission of Nigeria (ECN); Federal Institute of Industrial Research, Oshodi (FIIRO); National Agency for Science and Engineering Infrastructure (NASENI), Abuja; National Biotechnology Development Agency (NABDA); National Board for Technology Incubation (NBTI); National Centre for Technology Management (NACETEM); National Institute of Leather Science and Technology (NILEST), Zaria; National Office for Technology Acquisition and Promotion (NOTAP); National Research Institute for Chemical Technology (NARICT), Zaria; National Space Research & Development Agency (NARSDA); Nigerian Building and Road Research Institute (NBBRI); Nigerian Institute For Trypanosomiasis And Onchocerciasis (NITR); Nigerian Natural Medicine Development Agency (NNMDA); Project Development Institute (PRODA), Enugu; Raw Materials Research and Development Council (RMRDC), Abuja; Sheda Science and Technology Complex (SHESTCO), Abuja; and the Nigerian Institute of Science Laboratory Technology (NISLT).

2. National agencies/ Parastatals in the United Republic of Tanzania:

The Ministry of Industry and Trade (MIT) has two types of parastatals involved in supporting industries and businesses. Industrial support organizations include: The National Development Corporation (NDC),

Box 6.2: National agencies/parastatals with mandates to implement industrial and innovation policies *(cont.)*

which was established in 1962 to initiate, develop and guide the implementation of economically viable projects in partnership with the private sector; the Tanzania Industrial Research Development Organization (TIRDO), which was founded in 1979 to provide technical services to industries in the area of area of agro-technology and industrial chemistry, food and microbiology, energy and environment, information technology and instrumentation, leather and textile and materials science technology; the Tanzania Engineering and Manufacturing Design Organization (TEMDO), which was established in 1980 to promote engineering design, technology development and enhancement of the competitiveness of local manufacturing enterprises through provision of quality technical support services; the Centre for Agricultural Mechanization and Rural Technology (CAMARTEC), which was created in 1989 to disseminate improved technologies suitable for agricultural and rural development; The Export Processing Zone Authority (EPZA) which was set up in 2006 to coordinate, facilitate and promote investments in the export processing zones; and lastly the Small Industries Development Organization (SIDO), which was established in 1973 to develop the small industry sector. Business support organizations under MIT include: the Tanzania Trade Development Authority (TANTRADE), which was created in 1978 to spearhead Tanzania's Export endeavours; the College of Business Education (CBE) which was founded in 1965 to train highly competent and practice-oriented professionals; the Tanzania Bureau of Standards (TBS) which was established in 1977 to formulate standards and to undertake metrology quality control, testing and calibration and training.

The following organizations fall under the purview of the Ministry of Communication, Science and Technology (MST): the Dar es Salaam Institute of Technology (DIT); Mbeya University of Science &Technology (MUST); Nelson Mandela African Institute of Science &Technology (NM-AIST); Tanzania Telecommunications Company Ltd (TTCL); Tanzania Atomic Energy Commission; Tanzania Commission for Science and Technology (COSTECH); and The Tanzania Communications Regulatory Authority (TCRA). A number of institutions active in the areas of in agriculture and livestock, industry and energy, natural resources, medicine and public health, and universities and colleges are affiliated with MST.

3. National Agencies/ Parastatals in Ethiopia:

The Ministry of Industry oversees the following state institutions: The Privatization and Public Enterprises Supervising Agency, established in 2004 to implement the privatization programme and provide guidance and supervision to public enterprises; the Ethiopian Investment Agency, created in 1992 to promote private investment, primarily foreign direct investment; the Leather industry Development Institute (LIDI), set up in 2010 to facilitate the development and transfer of leather and leather products industries technologies and to enable the industries become competitive and experience rapid development; the Metal industry Development Institute, originally set up in 1973 but re-established in 2010 to facilitate the development and transfer of metals and engineering industries technologies, and to enable industries become competitive and beget rapid development; and the Ethiopia Kaizen Institute, established in 2011 to achieve greater effectiveness in the utilization of resources, quality improvements and enhanced performance capacity. Other parastatals under the Ministry of Industry include the: Textile Industry Development Institute; Food and Beverages and Pharmaceuticals Industry Development Institute (FB-PIDI); Chemical and Construction Input Industry Development Institute; and the Ethiopian Meat and Dairy Industry Development Institute.

The Ministry of Science and Technology has the following parastatals: The Science and Technology Information Centre, founded in 2011 to oversee the collection, selection, analysis and dissemination of science and technology information in Ethiopia; the Ethiopian Intellectual Property Office, established in 2003 to provide legal protection for intellectual property rights; the Ethiopian Conformity Assessment Enterprise, set up in 2011 to conduct inspection, laboratory testing and certification services to the public and to industry; the Ethiopian National Accreditation Office, created in 2011 and tasked to accredit the competence of Conformity Assessment Bodies, including testing laboratories involved with food and associated products, engineering and textiles; the Ethiopian Radiation Protection Authority, founded in 1993 to control and regulate the import, export, use, transport, dispose of any source of radiation; the Ethiopian Standards Agency, established in 2010 to oversee the development of standards, training and technical support on implementation of standards and to contribute to the country's economic and social development through technology transfer; and the National Metrology Institute of Ethiopia, created in 2011 to oversee the maintenance of Ethiopian National Measurement Standards and Certified Reference Materials and also to provide calibration, training and consultancy services in the areas of metrology and scientific equipment.

Source: Compiled by UNCTAD through primary data and official agency websites.

The recommendations in this regard include:

i. Coordination hurdles need to be tackled at the level of agencies and organizational structures.

Oftentimes, when the implementation framework is not clear, newer agencies created by the policy strategy tend to compete or have overlapping mandates with existing S&T agencies, leading to confusion among private sector actors as to which schemes are available and how they can be accessed. This is linked to the point made in the previous section on the need to ensure that existing initiatives maintain continuity as well.

For example, in the case of Nigeria, STI incentive schemes include:

(i)   Promoting cross-border collaboration that enables STI transformation;

(ii) Motivating the youth to take up careers in S&T fields;

(iii) Providing technology support services and other incentives;

(iv) Promoting public and private enterprises to invest at least 5 per cent of their profits before tax to the National Research and Innovation Fund; and

(v)   Providing funding and other incentives for continuing education of women in STI.

But as box 6.2 shows, there is a proliferation of agencies that are expected to implement many of these mandates, including the Nigerian Competitiveness Council, the Nigerian Investment Promotion Agency and other S&T agencies. It is therefore critical to set out the implementation of the incentives associated with these policy processes at the sectoral level as well as more generally; this is particularly important as several other agencies have sectoral mandates, e.g. the Nigerian Natural Medicines Development Agency (NNMDA), and the Nigerian Agency for Biotechnology Development (NABDA), which also perform various innovation-related activities. The survey also found that

it is unclear as to which of these agencies has been tasked with the responsibilities contained in the STI policy, and how performance/innovation support is to be provided in the absence of clear funding of these agencies/policy programmes. As a result, coordination amongst these agencies is rather weak, and initiatives are often duplicated and accompanied by institutional rivalries, thereby limiting their success. Nigeria's STI policy was also not accompanied by sectoral STI plans, although the framework sets out elaborate strategies for several sectors.

In a similar way, the Ethiopian STI policy, given its focus on technology transfer to create capacities for incremental innovation, seeks to devise incentive schemes to reward firms in the manufacturing and services sectors firms that have shown high performance gains through technology transfer. Ethiopia also seeks to offer various incentives to medium and large enterprises to help adapt foreign technologies. In order to facilitate this, the policy seeks to "establish and implement an appropriate national Technology Capability Accumulation and Transfer (TeCAT) system".

To ensure the success of this policy tool, it is imperative that it is accompanied by a clear implementation mechanism. Currently, the national STI policy aims to set up the TeCAT system[3] and provides for a set of strategies on technology transfer,[4] but there may be a need to link these more clearly. The TeCAT is not supported by concrete outcomes that can help establish and monitor its functioning, such as projections of the number of firms to be supported to adopt new technologies, projected increase in medium- or high-technology exports and estimated productivity increase. In conceptualizing the TeCAT system some of the other important avenues of technology transfer have also been left out; technology transfer could include, for example, the encouragement of joint ventures to promote technological transfers and learning. However, the policy does not align domestic regulations and investment incentives (including the current minimum capital requirement of $60,000 for foreign partnerships), with the technology (soft and hard) needs of industry.

ii. Policy changes should be accompanied by clear and enlarged budgets and staffing of skilled employees to facilitate their implementation

In all the three countries, the country-level investigations showed that national STI policy offices are experiencing increasing workloads, in part due to extended mandates, and faced significant challenges in their funding, personnel and capacity to coordinate the extremely difficult policy hurdles they currently have to overcome. All country surveys showed that funding was not only a major constraint within the innovation system affecting the firms, but also affected the ability of governmental agencies to offer substantive innovation support. Lack of funding thwarted the provision of technology incubation, R&D services, testing and quality assurance services, laboratory personnel, and even the provision of human skills. In Ethiopia, newly established institutes were not focused on R&D due to shortages in funding, for example two universities (Addis Ababa S&T and the Adema S&T) have yet to be endowed with their own research labs and other scientific infrastructure (see also box 5.6, chapter V).

iii. Develop common time frames and goals between STI and industrial policies

A critical issue that is common in almost all of the three countries is that a large number of initiatives that are relevant for STI are located within the ministry responsible for industrial development or in related agencies. This calls for closer coordination over the mid- to longer-term as the large number of initiatives and incentives that tend to run parallel in STI and industrial policy frameworks are all equally important, particularly those related to: (a) providing support to industry and businesses; (b) coordination of implementation; (c) day-to-day difficulties in interacting with industry; and (d) promoting exports and exacerbating finance issues. The country chapters strengthen these findings further, and show that national S&T

plans, or regular evaluation plans that emphasize policy processes play a non-negligible role in getting agencies, policymakers and those benefitting from the policy incentives to sit together and discuss relevant implementation issues.

iv. Importance of high-level governance structure and coordination

Mechanisms to foster high-level coordination are perhaps needed as it is often at the ministerial level that coordination and communication fails. At least two out of the three countries surveyed in the report have established committees that are tasked with coordinating the implementation of STI policy frameworks and coordinating this with industrial development. The Prime Minister/ President of the country should normally chairs such committees to facilitate ministerial level engagement. This is very important not only for effective policy coordination but also for support and successful implementation of the policy.

v. Best practices can only serve as a guideline

Although countries seek to emulate the successful practices of other developing countries (particularly East Asian economies) within policy frameworks (e.g. Ethiopia), country studies show that the policy impact is dependent on the current capacities of the private and public sectors in the countries concerned.

vi. Contextualization is key to achieving results

Policy frameworks or incentives tend to succeed better when they are contextualized to local needs and local circumstances. For example, a focus on SMEs, especially with a specific emphasis on what can be done to promote technological upgrading or expansion of firm size, remains important (see section C.4).

vii. Take stock of duplicated measures

The report found that in the case of finance, or even industry support, a large number of incentives for stimulating demand, direct or indirect finance (such as R&D grants) or industry support, are to be found in both policy frameworks, but are not well implemented or coordinated. In other cases, very important policy incentives were often not contained in either policy frameworks. It is therefore important to take stock of existing policy schemes before implementing them, or during the course of implementing them, to reinforce and manage scarce resources usefully (also see next section).

3. Policy monitoring and evaluation mechanisms to ensure efficient use of existing resources

Monitoring and evaluation (M&E) mechanisms are relevant from a variety of perspectives. They not only enhance coordination efforts but also point to the lack of funding of various initiatives as part of the stocktaking process. They also ensure that funding issues are taken into consideration and reviewed over time to evaluate: (a) where is the current funding being used? (b) What are the funding gaps to implement the goals of industrial and STI policies? (c) How can the gap be financed? (d) What are the best ways to share risk and partner with industry to effect transformation? (e) How to best allocate existing resources, and into what agencies? (f) Can agencies be streamlined and better defined? These are some of the issues that should form a core part of the monitoring and evaluation exercise.

Monitoring and evaluation exercises aimed at ensuring that existing resources and agency strengths are put to good use will play a pivotal role in policy effectiveness.

In support of this point, the surveys and interviews showed that most funding given to agencies supporting innovation is often spent on recurring expenses related to staff maintenance and running costs, with little or no reserve for innovation support infrastructure. In the United Republic of Tanzania, for example, about 95.1 per cent of the sums allocated to agricultural R&D goes

into staff salaries or operating expenses, leaving only 4.9 per cent for capital investments in 2011. Similarly, staff salaries and operating expenses account for about 83.4 per cent and 71.8 per cent of agricultural R&D in Nigeria and Ethiopia, respectively.[5] Similarly, supporting staff account for about 29.3 per cent (2010), 33.6 per cent (2007) and 37.9 per cent (2010) of the R&D expenditure in the United Republic of Tanzania, Nigeria and Ethiopia, respectively. By way of comparison, the share of support staff in relation to R&D personnel is smaller in other developed countries, e.g. Germany (16.8 per cent in 2011) and Japan (16.2 per cent in 2011), as well as in other developing countries with highly sophisticated R&D system, e.g. Hong Kong, China (5.5 per cent in 2010).[6]

In order to address these issues, the following recommendations could be considered:

a. Conceptualize monitoring from the start of the policy process

The report found that there is a concrete link between policy formulation, articulation of the monitoring process and clear delineation of funding. When policy goals are set, funding should be allocated accordingly, along with clear milestones for implementation and reporting to the monitoring bodies.

b. Ensure monitoring and regular follow-up

The report found that countries are trying to put in place monitoring mechanisms. For example, in Ethiopia a committee chaired by the Prime Minister meets every six months to review progress and impediments and is comprised of the National Science, Technology and Innovation Council (NSTIC), the Ministry of Science and Technology (MoST), other related ministries and the broader innovation support and research system. Going forward, it would be important to agree on the scope and methodology of how these monitoring systems will operate along with open assessments of budgets and assistance offered by various agencies

c. Monitoring should be based on institutional memory

The historical trajectories of countries are important. This point has been made time and again in the context of innovation studies when emphasizing the relevance of historical path-dependencies in the way policies evolve. This report explores these historical path-dependencies in the context of industrial and innovation policies, tracing the evolution of both frameworks from 1960s until the present day in the countries under study.

Very importantly, the chapters show that a lot of the issues related to policy failures and poor institutional performance can be traced back to past institutional failures. However, although familiar challenges abound, the interviews showed that few attempts have been made to assess and apply the learning of the country's own past as to why policies failed or what factors vitiated the policy processes. This is fundamental to experiment and derive successful coordination, and should be made a critical component of the monitoring and evaluation processes. For example, some of the goals of Ethiopia's current policy were already articulated in the 1980s, but they were not achieved. It would be important to assess the institutional reasons that hindered the policy in the 1980s and 1990s and to examine how to avoid repeating these errors in the future. Similarly, in Nigeria, the 2011 STI policy was informed by lessons drawn from the 1986 and 1997 S&T policies, as well as the 2003 policy. But the institutional memories of lack of coordination need to be actively tackled through regular interventions and the monitoring and evaluation process.

d. Financial realities are crucial

Policy effectiveness is largely decided by the resources that are allocated to support their implementation. The national STI and industrial development policies are often extremely ambitious, and seek to cover ground without financial allocation required for effectiveness. Foreign loans, domestic budgets and aid are often not disbursed on time and lead to delays or failures in the policy implementation process. For example, in Ethiopia, despite numerous challenges, the policy framework enjoyed widespread political and general support and a heady target of investing 0.6 per cent of the GDP into R&D was achieved in 2014.[7] However, programmes risk not being implemented in the absence of such continued financial support and expanded programme budgets.

4. Coordinate policymaking, governmental interventions and the business environment more closely

An important finding of this report is that policy is often out of synch with reality. That is, as opposed to the practical structure of local industry (often, in large part, mostly comprised of SMEs and the informal sector), industrial and innovation policy elaborate sectors of importance that are entirely high-tech, or require an institutional infrastructure that is very disconnected from the on-the-ground realities that firms have to face in their day-to-day activities. As summarized in section B.3 of this chapter, even in the so-called high-technology sectors, a number of the local firms are operating on the fringes of technological development. In the ICTs sector, many companies simply act as call centres or internet access to users (as opposed to any production or process improvements), while in the pharmaceutical sectors, many companies only distribute already packaged medicines or engage in traditional medicine-based preparations of low technological nature.

This is not to say that governments should not seek to promote sectoral priorities in high-technology sectors, such as space or satellite technologies, or even pharmaceuticals, electronics. But rather that industrial and innovation policies should pick activities that are promising and capable of being upgraded technologically with realistic prospects, as well as new sectors that

Figure 6.1:   Domestic credit to the private sector (as a percentage of GDP) in select countries and regions

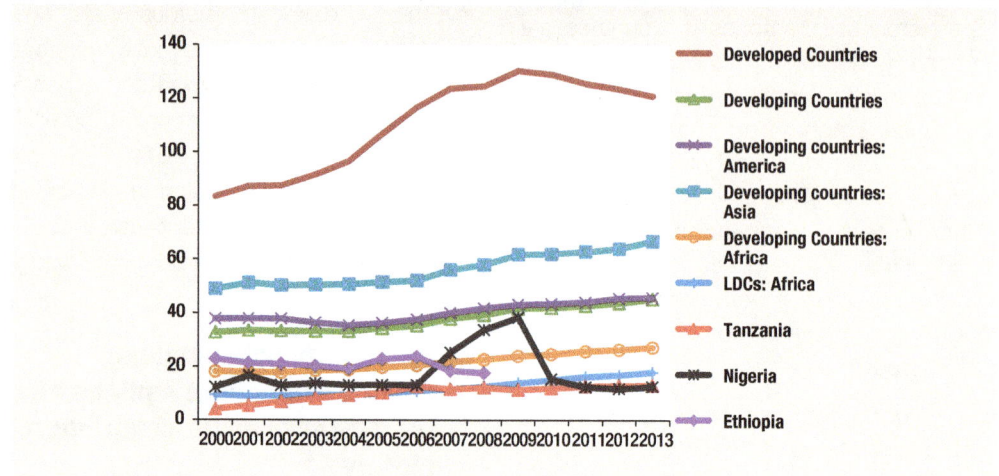

Source: *UNCTAD calculations based on WB WDI (accessed 20 October 2015)*

may be scaled up over time (Aiginger and Böheim, 2015).

To achieve the former, it is important incorporate a private sector perspective in the policy focus and the realm of policy discourse in the countries. The STI and industry policy frameworks should be adequately accompanied by both business and industry support organizations, which provide incentives for local firms, such as R&D grants and loans, tax credits and governmental procurement, all of which have met with much success in other developing countries. Business environments also need to be supported through business facilitation and enterprise incubation programmes. Regulatory measures that help business connect, interact and expand will be critical moving ahead, apart from enabling new means of financing business. In fact, one of the key issues that was raised in the country studies had to do with the way the issue of finance was managed. As figure 6.1 shows, financing remains a key issue for African countries, even in comparison with other country or regional groupings.

Thailand is an example of a country that uses policy mechanisms, such as government procurement as incentive for innovation.[8] However, there were policy implementation gaps on the question of innovation

finance in each of the three countries studied in this report.

In this regard, where there is a prevalence of small and medium scale enterprises, mid-term credit schemes for industry, guarantee schemes or micro-credit schemes constitute important external sources of finance for firms. These policy instruments need to be implemented to help SMEs meet the minimum requirements by banks (in terms of collaterals and sound business proposals) to access loans for innovative ventures. In addition to these, governments could facilitate the access of firms to venture capital and create an enabling environment for angel investments in strategic sectors. In the future, there may also be a need for newer instruments such as guarantee funds, with dedicated emphasis on venture capital, to promote niche areas of innovative activities within sectors.

***

It is becoming widely acknowledged that sustainable development rests more broadly on stable industrial development of a kind that can deliver better livelihoods to the people and eradicate poverty, as several goals of the recently adopted 2030 Agenda for Sustainable Development emphasize. In particular, Goal 9 encapsulates the dual objectives of promoting inclusive and sustainable industrialization and fostering innovation.

Almost all countries in the African region, and more widely in the developing world, including the three countries that were studied in depth for this report, are currently at a policy and developmental stage where industrial development through technological change *should be* a central, if not the most important, priority. Not only is there a policy transition towards that end, the field surveys were testimonies to the extent of political commitment to enacting elaborate industrial policy frameworks, and revising their S&T policies towards policies dedicated to innovation. Thirdly, the private sector in the African region (particularly in sub-Saharan Africa) is in dire need of greater support, and that enterprise policies are currently the weak link.

NOTES

1. http://www.tccia.com/tcciaweb/SMEtoolkit/introduction.htm

2. http://www.gdn.int/admin/uploads/editor/files/PPT_financing%20SME_FREDU.pdf

3. Ethiopia's 2012 STI Policy, Article 2.3, p. 4.

4. The strategies are: (a) import effective and appropriate foreign technologies and create capabilities of adaptation and utilization of these technologies in manufacturing and service providing enterprises; (b)a system to search, select, adapt, utilize as well as dispose imported technologies should be established and implemented; (c) establish and implement a system to use foreign direct investment (FDI) and other ways of supporting technology transfer; (d) strengthen technology transfer among and between various manufacturing and service providing enterprises; and (e) strengthen wide use of intellectual propriety, standards and other related information in support of technology transfer.

5. ASTI website (http://www.asti.cgiar.org/countries) accessed on 27 April 2015.

6. UNESCO Institute for Statistics database (http://data.uis.unesco.org/) accessed on 27 April 2015. Full time equivalent (FTE) figures were used.

7. Ministry of Science and Technology (2014). Ethiopia National Science, Technology and Innovation Policy, Addis Ababa.

8. See UNCTAD, Promoting Innovation Policies for Industrial Development in Thailand, forthcoming.

REFERENCES

Abeson F and Taku M (2007). "The perceived intervention effect of Nigerian government 1972 indigenization decree on US multinational corporations: A historical perspective". *Competition Forum*. 5(1): 151-155.

AfDB (2002). *Ethiopia - Structural Adjustment Programme - Completion Report*. Addis Ababa. Available at: http://www.afdb.org/fileadmin/uploads/afdb/Documents/Project-and-Operations/ADF-BD-IF-98-07-EN-3670048.PDF

AfDB (2010). "Ethiopia's economic growth performance: Current situation and challenges". *AfDB Economic Brief* 1(5), 17 September.

AfDB, OECD, UNDP and UNECA (2013). African Economic Outlook 2013. OECD Publishing, Paris.

Agboli M and Ukaegbu CC (2006). "Business environment and entrepreneurial activity in Nigeria: Implications for industrial development". *The Journal of Modern African Studies*, 44(1): 1-30.

Aghion P, Dewatripont M, Du L, Harrison A, Legros P. (2012). Industrial Policy and Competition. Available at: http://scholar.harvard.edu/files/aghion/files/industrial_policy_and_competition.pdf

Agosin M, Larrain C and Grau N (2010). "Industrial policy in Chile". *IDB Working Paper Series* No. IDB-WP-170.

Aiginger, Karl (2014) Industrial policy for a sustainable growth path. WWWFOREUROPE Policy Paper no:13.

Aiginger, K and Böheim M (2015) Fostering sustainable economic growth by redefining competitiveness and industrial policy: Towards a systemic policy approach aligned with beyond-GDP goals. *WIFO Policy Brief*, prepared for the UN Global Sustainable Development Report.

Altenburg T (2010). Industrial policy in Ethiopia. *DIE Discussion Paper* 4/2010. Bonn: Deutsches Institut für Entwicklungspolitik.

Amirahmadi H and Wu W (1994). *Export Processing Zones in Asia*. Center for Urban Policy Research, Rutgers, the State University of New Jersey.

Amsden A (1989). *Asia's Next Giant: South Korea and Late Industrialization*. Oxford: Oxford University Press.

Amsden A (2001). *The Rise of "The Rest": Challenges to the West from Late-Industrializing Economies*. Oxford: Oxford University Press.

Amsden A and Chu WW (2003). *Taiwan's Upgrading Policies*. Cambridge: The MIT Press.

Archibugi D and Michie J (2002). "Technical change, growth and trade: New departures in institutional economics". *Journal of Economic Surveys*, 12(3): 313-332.

Aregawi M (2005). Ethiopia's Economic Policy: 1950s-1970s. Available at: http://esai.org/myESAi/viewtopic.php?t=7155&sid=dc1bdacdfed39ad75a7cf5a35cd5b955

Assefa B (2009). Business Process Re-engineering in Ethiopia. Available at: http://www.grips.ac.jp/forum/af-growth/support_ethiopia/document/May09_berihu_bpr.pdf

Bank W and Madani D (1999). *A Review of the Role and Impact of Export Processing Zones*. World Bank Publications.

Benjamin N and Mbaye AA (2012). *The informal sector in Francophone Africa: Firm size, productivity and institutions*. Paris: Agence Francaise de Développement, Paris and Washington D.C: the World Bank.

Bianchi P and Labory S (2008). *International Handbook on Industrial Policy*. Edward Elgar Publishing.

Bairoch, P. and R. Kozul-Wright (1996).Globalization Myths: Some Historical Reflections on Integration, Industrialization and Growth in the World Economy, UNCTAD Discussion Paper No. 113. Geneva: UNCTAD.

Brautigam D (1997). "Substituting for the State: Institutions and industrial development in Eastern Nigeria". *World Development*, 25(7): 1063-1080.

Briggs IN (2007). Nigeria: Mainstreaming trade policy into national development strategies. *ATPC Work in Progress No. 52*, Addis Abeba: UNECA.

Chang, Ha-Joon and R. Kozul-Wright (1994). "Organizing development: Comparing the systems of entrepreneurship in Sweden and South Korea", *The Journal of Development Studies*, 4, pp. 859-891

Chang, Ha-Joon (2001). Infant Industry Promotion in Historical Perspective. A Rope to Hang Oneself or a Ladder to Climb With? Document prepared for the conference on "Development Theory at the Threshold of the Twenty-first Century", Economic Commission for Latin America and the Caribbean (ECLAC), August.

Chang HJ (2011). *23 Things They Don't Tell You About Capitalism*. Bloomsbury USA, New York.

Chete LN, Adeoti JO, Adeyinka FM, Ogundele O (2014). Industrial development and growth in Nigeria: Lessons and challenges. *UNU-WIDER Working Paper No. 2014/19*.

Cimoli M, Dosi G and Stiglitz J (2009). *Industrial Policy and Development: The Political Economy of Capabilities and Accumulation*. Oxford, Oxford University Press.

Cissé, F, Choi, J.E and Maurel, M. (2014). "Scoping paper on industry in Senegal". *WIDER Working Paper 2014/157*. Available at: http://www.trademarksa.org/news/swaziland-commerce-ministry-wants-develop-industrial-policy

Dahlman C (2001). *China and the Knowledge Economy: Seizing the 21st century*. World Bank Publications. Washington D.C.: World Bank.

Daka E and Toivanen H (2014). "Innovation, the informal economy and development: The case of Zambia". *African Journal of Science, Technology, Innovation and Development*, 2014: 1–9.

Dervis K and Page JM Jr (1984). "Industrial Policy in Developing Countries", *Journal of Comparative Economics*, Volume 8 (4): 436–451, December.

ESAURP (2012). *Transforming the Informal Sector: How to Overcome the Challenges*. TEMA Publishers Company, Dar es Salaam: United Republic of Tanzania.

Etzkowitz H and Roest M (2008). *Transforming University-Industry-Government Relations in Ethiopia*. IKED, Sweden Available at: http://www.aau.org/sites/default/files/urg/docs/trans_univ_ind_gov_rel_ethiopia.pdf

Evans, Peter (1995) *Embedded autonomy: States and industrial transformations*. Princeton, Princeton University.

Foray, D. (2015). Smart Specialisation - Opportunities and Challenges for Regional Innovation Policy. Regions and Cities Routledge. EPFL-BOOK-201601.

Fukasaku K, Kawai M, Plummer MG and Trzeciak-Duval A (2005). Policy Coherence Towards East Asia Development. Development Challenges for OECD Countries. OECD Publishing, Paris.

Gebreeyesus M (2013). "Industrial policy and development in Ethiopia: Evolution and present experimentation". *UNU-WIDER Working Paper No. 2013/125*.

Gray H (2013). "Industrial policy and the political settlement in Tanzania: Aspects of continuity and change since independence". *Review of African Political Economy*, 40 (136): 185-201.

Georghiou, L., Li, Y., Uyarra, and Edler J. (2014). "Policy instruments for public procurement of innovation: Choice, design and assessment". *Technological Forecasting and Social Change*, Volume 86: 1-12.

Hausmann R, Hwang J and Rodrik D (2005). "What you export matters". *NBER Working Paper no. 11905*.

Helleiner G (1976) *A World Divided: The Less Developed Countries in the International Economy*. Cambridge: Cambridge University Press.

Hidalgo C et al. (2007). "The product space conditions the development of nations". *Science Magazine*. 317 (5837): 482–487.

IDB (2010). *The Age of Productivity – Transforming Economies from the Bottom Up*. London and New York: Palgrave.

ILO, UNIDO and UNDP (2002). *Roadmap Study for the Informal Sector in Tanzania*. Dar es Salaam, United Republic of Tanzania.

Imbs J and Wacziarg R (2003). "Stages of Diversification". *American Economic Review*, 93(1): 63-86.

Imevbore AMA (2001). Industry and sustainable development in Nigeria – achievements and prospects. In Luken R, Alvarez J, Hesp P, (eds), *Developing Countries Industrial Source Book*, p. 64-69. Vienna: UNIDO.

IMF (2004). T*he Federal Democratic Republic of Ethiopia: Poverty Reduction Strategy Paper Annual Progress Report*. Country Report No. 04/37.

InnovaChile (2010). *Chile: An innovation hub in Latin America*. InnovaChile CORFO.

Johann M and Nelius B (2007). *African Science and Technology Profiles: Science and Technology in the Federal Republic of Ethiopia*. Study commissioned by the Department of Science and Technology. Centre for Research on Science and Technology, Stellenbosch University.

Johnson C (1986). "The institutional foundations of Japan's industrial policy". In Barfield CE and Schambra WA (eds), *The Politics of Industrial Policy,* American Enterprise Institute, p. 187-205.

Johnson C (1999). The Developmental State: Odyssey of a Concept. in Woo-Cummings, M (ed), 1999, *The Developmental State*, Cornell: Cornell University Press.

Johansson H and Nilsson L (1997). Export processing zones as catalysts. *World Development*. 25(12):2115–2128.

Kapunda SM (2014). Trade, industrial policy and development in the era of globalization in Africa: The case of Botswana and Tanzania. in Moyo T, (ed.), *Trade and Industrial Development in Africa: Rethinking Strategy and Policy*. Dakar, CODESRIA. Ch. 3: 47-66

King, W.R. (1991). *Ensuring Quality Science and Technology Education: The role of agencies outside the school*. Proceedings of the CASTME/COL North American Regional Seminar on Quality in Science, Technology, and Mathematics Education, University of British Colombia, Vancouver, Canada, 15-19 April, 44-53.

Lall S (1990). *Building industrial competitiveness in developing countries*. Paris, France: Development Centre of the Organisation for Economic Co-operation and Development.

Lall S, Navaretti B, Teitel GS and Wignaraja G (1994). *Technology and enterprise development: Ghana under structural adjustment*. Basingstoke: Macmillan.

Lall S and Teubal M (1998). "Market-stimulating" technology policies in developing countries: A framework with examples from East Asia. *World Development*, Elsevier, vol. 26(8): 1369-1385.

Lawson A, Booth D, Msuya M, Wangwe S and Williamson T (2005). *Does General Budget Support Work? Evidence from Tanzania*. London and Dar es Salaam, Overseas Development Institute and Daima Associates.

Lee, Keun (2013). Schumpeterian Analysis of Economic Catch-up: Knowledge, Path-Creation, and the Middle-Income Trap, Cambridge University Press.

Lin, JY (2014) Growth identification and facilitation: The new industrial policy for inclusive and sustainable industrial development. UNIDO PowerPoint Presentation.

Lin JY and Chang HJ (2009). Should industrial policy in developing countries conform to comparative advantage or defy it? A debate between Justin Lin and Ha-Joon Chang, *Development Policy Review*, 27(5): 483-502.

Lundvall BÅ (1993). User-producer relationships, national systems of innovation and internationalisation, in Foray, D. and Freeman, C. (eds.): *Technology and the Wealth of Nations*, Pinter Publishers.

Malmberg A and Maskell P (2002). "The elusive concept of localization economies: towards a knowledge-based theory of spatial clustering". *Environment and Planning A*. 34(3): 429 – 449.

Maskus K (2014). *Benefits and Costs of the Science and Technology Targets of the Post-2015 Agenda*, Copenhagen

Consensus Center Publication.

Mazzucato M (2013). *The Entrepreneurial State: Debunking Public vs. Private Sector Myths*. London, Anthem Press.

McCormick, D. (1999). "African Enterprise Clusters and Industrialization: Theory and Reality". *World Development*, 27(9), 1531-1551.

McIntyre J, Narula R, and Trevino. (1996). "The role of export processing zones for host countries and multinationals: A mutually beneficial relationship?" *The International Trade Journal* 10 (4): 435-466.

McMillan MS and Rodrik D (2011). "Globalization, structural change and productivity growth". *NBER Working Paper 17143*.

Meyer-Stamer, J. (2009): Moderne Industriepolitik oder postmoderne Industriepolitiken? (Modern Industrial Policy or Postmodern Industrial Policies?) *Schriftenreihe Moderne Industriepolitik 1/2009*. Berlin: Friedrich-Ebert-Stiftung.

Ministry of Trade and Industry – United Republic of Tanzania (2012). National Baseline Survey Report for Micro Small and Medium Enterprises in Tanzania. Ministry of Trade and Industry.

MoFED Ethiopia (2010). *Ethiopia's Growth and Transformation Plan: At-a-Glance*. Addis Ababa. Available at: http://photos.state.gov/libraries/ethiopia/427391/PDFpercent20files/GTPper cent20At-A-Glance.pdf

Moudud J (2010). *Strategic Competition, Dynamics, and the Role of the State: A New Perspective*. New Directions in Modern Economics Series, Edward Elgar Press.

Moudud J (2011). "Constrained autonomy and the development state: From successful developmentalism to catastrophic failure". International Affairs at the New School, *International Affairs Working Papers 2011-10*.

National Planning Commission of Nigeria (2010). 2010 Performance Monitoring & Evaluation Report – For Federal Ministries Departments and Agencies. Abuja, National Planning Commission.

National Bureau of Statistics- United Republic of Tanzania (2013). Quarterly Production of Industrial Commodities: 2004-2012 - Tanzania Mainland, National Bureau of Statistics and Ministry and Finance, Dar es Salaam.

National Bureau of Statistics- United Republic of Tanzania (2014). Revised National Accounts Estimates for Tanzania Mainland - Base Year, 2007. National Bureau of Statistics and Ministry of Finance, Dar es Salaam.

Naudé W (2010). "New Challenges for Industrial Policy". *UNU-WIDER Working Paper No. WP/107*.

Nayyar D (2012). The MDGs after 2015: Some reflections on the possibilities. UN System Task Team on the Post-2015 UN Development Agenda. Available at: http://www.un.org/millenniumgoals/pdf/deepak_nayyar_Aug.pdf

Nelson RR (1987). *Understanding Technical Change as an Evolutionary Process.* Amsterdam: North-Holland Press.

Nelson, R. and Pack, H. (1999). "The Asian Miracle and Modern Growth Theory," *Economic Journal*, Royal Economic Society, vol. 109(457): 416-36.

NEPAD (2010) African Innovation Outlook, 2010. Pretoria, AU-NEPAD. Available at: http://www.nepad.org/system/files/June2011_NEPAD_AIO_2010_English.pdf.

Njoku A and Ihugba O (2011). *Unemployment and Nigerian Economic Growth (1985-2009)*. Alvan Ikoku Federal College of Education, Owerri Imo State, Nigeria. Available at: http://www.hrmars.com/admin/pics/371.pdf

Nyerere, Julius, K. (1977). *Ujamaa: Essays on Socialism*. Dar es Salaam and Nairobi: Oxford University Press.

OECD (2010). *Perspectives on Global Development: Shifting Wealth*. Paris: OECD Publishing.

OECD (2014), *OECD Science, Technology and Industry Outlook 2014*. Paris: OECD Publishing. Available at : http://dx.doi.org/10.1787/sti_outlook-2014-en

Osoro NE (2009). *Domestic resource mobilization – Tanzania: A case study*. North-South Institute.

Otsuka K (2012). "University patenting and knowledge spillover in Japan: Panel-data analysis with citation data". *Applied Economic Letters*, 19(11): 1045-1049.

Oyelaran-Oyeyinka B (1997a). "Technological learning in African industry: A study of engineering firms in Nigeria". *Science and Public Policy*, 24(5): 309-318.

Oyelaran-Oyeyinka B (1997b). *Nnewi: An Emergent Industrial Cluster in Nigeria*. Ibadan, Technopol Publishers.

Oyelaran-Oyeyinka B (1998). Technological capability building in the South: Lessons and opportunities for sub-Saharan Africa. Addis Abeba: UNECA.

Oyelaran-Oyeyinka B and Gehl Sampath P (2007). *Innovation in African Development: Case studies of Uganda, Tanzania and Kenya*. Washington, D.C.: World Bank:

Oyelaran-Oyeyinka, B (2014) *Rich Country Poor People; Nigeria's Story of Poverty in the Midst of Plenty*. Ibadan: Technopol Press.

Pack H and Saggi K (2006). "The case for industrial policy: A critical survey". *World Bank Research Policy Working Paper* no 3839.

Page J (2014). Three myths about African industry. in *Foresight Africa*. Washington D.C.: Africa Growth Initiative, Brookings.

Poynter TA (1982). "Government intervention in less developed countries: The experience of multinational companies". *Journal of International Business Studies*, 13(1): 19-25.

Poynter TA (1986). "Political risk: Managing government intervention". *Columbia Journal of World Business*, winter.

Primi A (2013). The Return of Industrial Policy: (What) Can Africa learn from Latin America? Working paper prepared for JICA/IPD Africa Task Force Meeting Yokohama, Japan.

Reinert, Erik S (2007) *How Rich Countries Got Rich and Why Poor Countries Stay Poor*. London: Constable.

Robinson JA (2010). Industrial policy and development: A political economy perspective. In Lin JY and Pleskovic B (eds.) *Lessons from East Asia and the global financial crisis*. Annual World Bank conference on development Economics Global. Washington, D.C.: World Bank.

Rodrik, Dani (1988). "Industrial organization and product quality: Evidence from South Korean and Taiwanese exports". *NBER Working Paper no: 2722*.

Rodrik, Dani (2004). "Industrial policy for the twenty-first century". *CEPR Discussion Paper DP4767*.

Rodrik, Dani (2007). *One Economics, Many Recipes: Globalization, Institutions, and Economic Growth*, New Jersey: Princeton University Press.

Rodrik, Dani (2013). "Unconditional Convergence in Manufacturing". *Quarterly Journal of Economics*, 128 (1), 165-204.

Rodrik, Dani (2014). "An African Growth Miracle?" *NBER Working Paper 20188*.

Sapelli C (2003). "The political economics of import substitution industrialization". *Documento de Trabajo No. 257*, Pontificia Universidad Catolica de Chile.

Schumpeter JA (1934). The nature and necessity of a price system, in Harris SE, Bernstein EM (eds) *Economic reconstruction*, New York: McGraw-Hill.

Schumpeter JA (1942). *Capitalism, Socialism and Democracy* (2nd ed.). Floyd, Virginia: Impact Books (2014 Reprint).

Stiglitz J and Greenwald BC (2014). *Creating a Learning Society: A New Approach to Growth, Development, and Social Progress*. New York: Columbia University Press.

Tefera A and Tefera T (2013). Coffee Annual Report 2013. Report Number: ET-1302. USDA Foreign Agricultural Service. Global Agricultural Information Network. Available at: http://gain.fas.usda.gov/Recent%20GAIN%20Publications/Coffee%20Annual_Addis%20Ababa_Ethiopia_6-4-2013.pdf

Tema BO and Mlawa H (2009). *Assessment Report on Integration of Science, Technology and Innovation in Mkukuta and Mkuza*. Dar es Salaam, United Republic of Tanzania: UNESCO.

The Economist (2013). Mining in Chile: Copper solution. From the print edition: Business. Available at http://www.economist.com/node/21576714/print.

Turner R (2011). "Evaluation of Leadership and Management Advisory Service. Department for Business", *Innovation and Skill Research paper number 5*. Available at:

https://www.gov.uk/government/uploads/system/uploads/attachment_data/file/32300/11-1301-evaluation-leadership-and-management-advisory-service.pdf.

UNCTAD (2003). *Africa's Technology Gap. Case Studies on Kenya, Ghana, Uganda and Tanzania*. New York and Geneva: United Nations.

UNCTAD (2004). *The Least Developed Countries Report 2004: Linking International Trade*

with Poverty Reduction. New York and Geneva: United Nations.

UNCTAD (2006). *The Least Developed Countries Report 2006: Developing Productive Capacities*. New York and Geneva: United Nations.

UNCTAD (2010). *The Least Developed Countries Report: Towards a New International Development Architecture for LDCs*. New York and Geneva: United Nations.

UNCTAD (2012). *Technology and Innovation Report – Innovation, Technology and South-South Collaboration*. New York and Geneva: United Nations.

UNCTAD (2013). *World Investment Report 2013- Global Value Chains: Investment and Trade for Development*. New York and Geneva: United Nations.

UNCTAD (2015). *Fron Decisions to Actions.* Report of the Secretary-General of UNCTAD to the fourteenth session of the United Nations Conference on Trade and Development. New York and Geneva: United Nations.

UNCTAD (forthcoming), *Promoting Innovation Policies for Industrial Development in Thailand*. New York and Geneva: United Nations.

UNESCO (2009). S&T Policy Structure of Ethiopia. UNESCO Workshop Presentation. Mombasa, Kenya.

UNESCO (2010). *UNESCO Science Report 2010*. Paris, UNESCO. Available at: http://unesdoc.unesco.org/images/0018/001899/189958e.pdf

UNIDO (2013). *Industrial Development Report 2013*. UNIDO, Vienna.

UNIDO and Government of the United Republic of Tanzania (2012). *Tanzania Industrial Competitiveness Report 2012*. Vienna: United Nations.

UN System Task Team on the Post-2015 UN Development Agenda (2012). *Realizing the future we want for all: Report to the Secretary-General*. New York: United Nations.

UNU-MERIT (2015). "Innovation for Development in Southern & Eastern Africa: Challenges for Promoting ST&I Policy". *Policy Brief Number 1, 2015.*

Wangwe S (2013). *Learning to compete: Industrial development in Tanzania*. Presentation at L2C UNU-WIDER Conference, 24-25 June, Helsinki.

Wangwe S, Mmari D, Aikaeli J, Rutatina N, Mboghoina T and Kinyondo A (2014). "The performance of the manufacturing sector in Tanzania – Challenges and the way forward". *UNU-WIDER Working Paper* no. 2014/085.

Wangwe S and Rweyemamu D (2004). *Challenges towards promoting industrial development in Tanzania*. Paper presented at a workshop to mark the 10th Anniversary of ESRF – 17 December 2004.

Warwick K (2013). "Beyond industrial policy: Emerging issues and new trends", *OECD Science, Technology and In-*

dustry Policy Papers, no. 2. Paris: OECD Publishing.

Watson P. (2001). "Export Processing Zones: Has Africa Missed the Boat? Not yet!" *African Region Working Paper Series No. 17*. Washington D.C: World Bank.

World Bank (2011). *Fostering Technology Absorption in Southern African Enterprises*. Washington D.C: World Bank.

World Bank (2012). *The informal sector in Francophone Africa: firm size, productivity and institutions*. Washington D.C: World Bank.